To Know Is to Compare

Studying Social Media across Nations,
Media, and Platforms

Mora Matassi and Pablo J. Boczkowski

The MIT Press

Cambridge, Massachusetts | London, England

The MIT Press would like to thank the anonymous peer reviewers who provided comments on drafts of this book. The generous work of academic experts is essential for establishing the authority and quality of our publications. We acknowledge with gratitude the contributions of these otherwise uncredited readers.

This book was set in Stone Serif and Stone Sans by Westchester Publishing Services. Printed and bound in the United States of America.

Library of Congress Cataloging-in-Publication Data

Names: Matassi, Mora, author. | Boczkowski, Pablo J., author.
Title: To know is to compare : studying social media across nations, media, and platforms / Mora Matassi and Pablo J. Boczkowski.
Description: Cambridge, Massachusetts : The MIT Press, [2023]. | Includes bibliographical references and index.
Identifiers: LCCN 2022030583 (print) | LCCN 2022030584 (ebook) | ISBN 9780262545938 (paperback) | ISBN 9780262374989 (epub) | ISBN 9780262374972 (pdf)
Subjects: LCSH: Online social networks—Study and teaching. | Social media—Study and teaching.
Classification: LCC HM742 .M365 2023 (print) | LCC HM742 (ebook) | DDC 302.231—dc23/eng/20220707
LC record available at https://lccn.loc.gov/2022030583
LC ebook record available at https://lccn.loc.gov/2022030584

10 9 8 7 6 5 4 3 2 1

Contents

Contents

Acknowledgments

The story of this book is a play in two acts.

Act one took place in June 2018 during a meeting at the Luc-ciano's gelateria branch near to the Google offices in Buenos Aires, Argentina. Among other topics, during that conversation we talked about our shared frustration with the relative lack of comparative perspectives in scholarship on social media. We thought that it could be worth exploring the possibility to develop a theoretical approach that would put comparative work at the center. Mora was about to start her doctorate at Northwestern University with Pablo as her advisor. For well over a year, we met weekly to discuss ideas and texts and ended up writing a manuscript that we submitted to a journal in fall 2019.

Act two occurred in April 2020 during a long WhatsApp chat dis-cussing how to approach yet another round of revisions of the jour-nal article. Mora had returned to Buenos Aires to spend what at the time seemed like just a few months of lockdown in the company of her family. As the conversation progressed, it became increasingly evident to both of us that the journal article format, with its strict word limit and genre expectations, was constraining our ability to both fully develop the various facets of our argument and do so in ways that would show not only its theoretical import but also its social relevance. We ultimately realized that a book would provide a better canvas to lay out our vision, and so we wrote a proposal, secured a contract, and embarked in a year-long writing journey.

The writing process and our weekly calls about it became a lifeline as the global shutdown continued far beyond what we had initially imagined and the pandemic's disruption reached far beyond the imaginable. The book in your hands is the result of an intellectual ride that helped us to hold onto something stable and to feel alive while many of the things we knew were losing stability and the level of loss was becoming staggering.

We were not alone during this creative and writing journey. Eugenia Mitchelstein, Ignacio Siles, Facundo Suenzo, and Celeste Wagner took the time to carefully comment on different versions of the book manuscript and provided us with bright and eye-opening perspectives. Larry Gross and anonymous reviewers contributed valuable feedback on earlier versions of some of the ideas further developed in this book and which were published in Matassi and Boczkowski (2021). Eszter Hargittai and Henry Jenkins, as well as David Park, Steve Jones, Santiago Marino, and anonymous journal article reviewers for *New Media & Society* offered helpful suggestions. Jack Bandy, Diego Gómez-Zara, Nicholas Hagar, Chelsea Peterson-Salahuddin, Daniel Trielli, and Erique Zhang provided important insights during a doctoral seminar on "Media Meet Technology" that Pablo taught at Northwestern University during the winter 2019 quarter, which Mora also took, and in which we first assessed the potential of the ideas that animate this book. Organizers and attendees at the Comparative Research in Media Studies Workshop, organized by Cicant—Lusófona University, University of Oslo, University of Bergen, and Catolica Research Centre for Psychological, Family and Social Wellbeing, and in which Mora presented ideas contained in this book, asked us truly useful questions.

Working with MIT Press has been wonderful. Gita Manaktala, our editor, championed the project from day one, secured terrific reviews, gave us excellent insights, and was always available to answer any of our queries. At different stages in the project Erika Barrios and Suraiya Jetha provided excellent assistance. A book can unfortunately have only one home, but we also want to acknowledge the interest, support, and contributions of Mary Savigar at Polity Press. Finally, we want to thank the anonymous reviewers at MIT Press and Polity for their most useful feedback.

Pablo wholeheartedly thanks his friends and family for their support throughout the writing process during a most challenging time. As the saying goes, it takes a village, and he feels blessed for the one he inhabits. He is also deeply grateful to Mora for leading the writing process with unparalleled creativity, energy, commitment, and steadiness. Reaching middle age and a stage of career maturity has made him cherish the critical value of intellectual partnerships that make learning a lifelong process marked by a sense of excitement and discovery. Since this is the second time that he coauthors a book with a student while they are in graduate school—the first was with Eugenia Mitchelstein for *The News Gap* almost a decade ago, also published with MIT Press—he was aware from the start of this project of the unique intellectual energy that emerges when two very different generational outlooks coalesce. This generational encounter has marvelously shaped both what we say and how we chose to say it in this book, and he could not think of a better writing partner than Mora for this endeavor.

Mora is grateful to her family, friends, and colleagues for their inspiring presence in her life. She specially thanks her mentor and coauthor Pablo for opening up a world of opportunities and fostering each of them with care and generosity. Whenever the goals ahead seem too difficult, Pablo is there to offer an encouraging word and a clear path forward. During the process of writing this book, he provided invaluable conditions for intellectual growth and kind collaboration—putting humanity always first. Mora also specially thanks her parents, Claudia and Víctor. They taught her the actual meaning of the words *confianza, apoyo, alegría, y amor*. Throughout all of her personal and professional projects, they have tirelessly supported her, amplifying what helping and nurturing a human being can mean. She will be forever grateful for the countless times in which they patiently, brightly, and eagerly listened, advised, and cared for her happiness, across Monte Grande, Buenos Aires, Boston, and Chicago, and across thousands of *mates, cafés*, and meaningful conversations—offline and online. Mora's intellectual journey simply wouldn't exist if it weren't for Claudia and Víctor's time, infinite generosity, joy, and love. She dedicates this book to them.

Buenos Aires and Evanston, Illinois, April 29, 2022

1

Nations, Media, and Platforms

Introduction

For many of us it is no longer easy to remember how it was to live in a world without social media. How did we share our daily existence, find out what others were up to, get breaking news, watch DIY videos, play games, or even kill time before Facebook, WhatsApp, Instagram, WeChat, Twitter, YouTube, TikTok, Snapchat, and Twitch, among other platforms, became part of our vernacular? It is often hard to believe that only a few years ago our everyday communication practices were greatly different from what they are nowadays, to the extent that sometimes the early 2000s feel more like the mid-twentieth century than a decade ago in our rearview mirror.

The normalization of social media has also reached the field of communication and media studies. This domain of inquiry, which during the previous century was marked by a concern with broadcast and print media, has recently been almost obsessed with all things platforms. For instance, a keyword search conducted in summer 2021 for "television," "newspapers," and "radio" included in the titles of papers published in communication between 2012 and 2021 yielded 2,272, 1,242, and 923 results on the database Web of

Science, respectively, for a total of 4,437 entries. The same search using "social media" had 4,103 entries, almost as many as the previous three traditional media keywords combined. Furthermore, while the trend for television shows a slow decline in recent years, that for social media exhibits a steep upward trend that has widened the distance between the two: there were almost three times more publications with "social media" than with "television" in their titles during 2020 (616 vs. 214).

The burgeoning scholarship on social media has made fundamental contributions about a broad range of critical issues (boyd and Ellison 2007; boyd 2014; Baym 2015; Burgess, Marwick, and Poell 2018)—covering a wide array of topics such as identity making and self-presentation (Donath and boyd 2004; Marwick and boyd 2011), relationship maintenance and social capital (Ellison, Steinfield, and Lampe 2007; Quan-Haase and Young 2010), and political participation and activism (Tufekci and Wilson 2012; Jackson and Foucault Welles 2015). Importantly, a critical strand of social media studies has increasingly shed light on three key and interconnected aspects structuring the production, distribution, and use of platforms—namely, dynamics of racial and ethnic discrimination (Nakamura and Chow-White 2012; Gillespie 2018; Noble 2018; Brock 2020), the platform economy of social media (Gillespie 2010; Fuchs 2016; Plantin et al. 2018; Nielsen and Ganter 2022), and the logics of datafication and algorithmic bias (Bucher 2012; van Dijck 2014; Crawford and Gillespie 2016; Roberts 2019).

However, beneath the diversity of contributions from this scholarship there are three sets of common limitations that have characterized most social media research to date. First, the majority of studies has examined empirical phenomena taking place within the confines of a single country—and often located in the Global North. Second, the bulk of the research has focused on social media without connecting them to dynamics affecting other media and communication technologies, especially traditional or predigital alternatives. Third, most scholarship has tended to concentrate on patterns related to a single platform at a time—usually either Facebook or Twitter. Taken together, these three limitations lead to a portrayal of the everyday realities of social media that is at best partial, and sometimes even distorted, relative to how platforms have been designed, distributed, and adopted. Let us briefly address each limitation separately.

First, platforms are deployed and used all over the world, with information flows connecting accounts located in multiple countries via reactions, comments, and shares. Single-country studies artificially cut this dense web of communication that does not necessarily stop at national borders—with some exceptions due most commonly to restrictions imposed by authoritarian regimes and/or inequalities in access. Furthermore, the Global North countries that have been the dominant foci of most scholarship collectively constitute only 14 percent of the planet's population. They also have distinguishing structural and cultural features that tend to be different from those that characterize the rest of the world.

Second, from the time of their development and throughout their meteoric rise to becoming a mainstay of contemporary communication practices, social media have entered a mediated communication landscape already featuring a rich array of artifacts and their associated cultures of production, circulation, and use. This broader landscape has shaped the brief but intense evolution of social media in at least two major ways. First, as a handful of studies has shown, prior communication technologies and genres—from the personal diaries of the nineteenth century to the reality television shows of the late twentieth century—have been precursors of what later became key aspects of the design and use of social media platforms (Marwick 2013; Hermida 2014; Humphreys 2018). Second, as platforms have become more popular, a significant portion of their use has been either concurrent with that of other media, for instance, in the increasingly common phenomenon of second screening (Doughty, Rowland, and Lawson 2012; Gil de Zúñiga, García-Perdomo, and

McGregor 2015; Gil de Zúñiga and Liu 2017), or in relation to content originally produced by other media such as social media activity around news stories, television shows, and movies (Highfield, Harrington, and Bruns 2013; Ksiazek, Peer, and Lessard 2016; Gutiérrez-Martín and Torrego-González 2018). Thus, a focus on social media that isolates them from the broader media and communication landscape effectively removes historical and contemporary connections that have variously shaped the everyday life of platforms.

Third, despite the single-platform focus of most scholarship, social media use is remarkably plural. To begin, according to DataReportal, "the typical user actively uses or visits an average of 7.5 different social media platforms each month."[1] Moreover, mounting scholarly evidence suggests that people use a particular platform in relation to how they use the others they routinely access (Boczkowski, Matassi, and Mitchelstein 2018; Tandoc Jr., Lou, and Min 2019). In addition, contrary to the overwhelming attention paid to Facebook and Twitter in the existent academic literature, there are dozens of other platforms that have garnered the interest of users. As of 2022, there are thirty platforms with at least 100 million users each, and while Facebook occupies the first place in this list with 2.9 billion monthly active users, Twitter is in the seventeenth place with less than 400 million.[2] The combination of a single-platform focus that is at odds with the greater plurality of the user experience, and a concentration on Facebook and Twitter in a field that includes a much wider array of alternatives about which we know comparatively much less, has unnecessarily diminished our accounts of the role of social media in the experiences of billions of people around the world.

Beneath these three sets of limitations lies a common denominator characterized by the absence of comparative analyses across nations and regions, across media, and across platforms. Thus, in this book we aim to contribute to scholarship on social media by developing original comparative perspectives that intend to overcome these limitations. Our contribution builds on studies that have already shown the value of comparing social media phenomena on at least one of the three dimensions highlighted above—across nations and regions, media, and platforms. But, unlike most of these studies, in which comparative work seems to have emerged as a by-product of trying to answer specific research questions, we propose to turn the practice of comparison into the epistemological principle framing our intellectual agenda. In this sense, our proposal is premised on the idea that foregrounding systematic comparative efforts across nations, media, and platforms holds great potential for social media scholarship.

Pursuing this intellectual agenda entails a stance which signals that at a very basic level—as the title of this book encapsulates—to know is to compare. Simply put, by this we mean that whatever it is that we are able to know, we do so as a result of contrasting two or more entities. We view this comparative stance as related to, but distinct from, issues of method and theory. On the one hand, its enactment can be undertaken utilizing a variety of methods, as we will illustrate with specific examples throughout the book. On the other hand, the answers to the questions animated by this stance are amenable to explanation by a range of theoretical frameworks, as we will also show in future chapters. In this sense, the core of our intellectual agenda is a broader epistemological umbrella that encompasses issues of methods (how to gather and analyze data) and theory (how to explain variance in the findings), which leads to a refiguration of, paraphrasing Clifford Geertz, "what is it that we want to know" (1980, 178).

As James R. Beniger argued three decades ago, "*all* social science research is comparative" (1992, 35; emphasis in the original). Comparative scholarship has a long history in communication and media studies, where it has been appreciated for different reasons: its capacity to attain generalizations from past periods and singular contexts, its power to test hypotheses, its ability to properly contextualize and thus avoid the naturalization of specific cases, and its promotion of international academic collaborations, among others (Esser and Hanitzsch 2012). Within communication and media studies, comparative research has been enacted in a variety of ways. The novelty of our proposal stems partly from foregrounding the role of comparisons coupled with the impetus, continuing with the landmark essay by Clifford Geertz (1980), to "blur genres" of comparative research in the service of accounts of social media phenomena that reflect their global, de-westernized, inter-media, and multiplatform existence. Thus, in the next section we continue our argumentative journey by elaborating on how different varieties of comparative research have informed this book.

Varieties of Comparative Research

Our analyses of cross-national and regional dynamics are indebted to a long tradition of scholarship in communication and media studies that has used almost interchangeably the terms *comparative* and *international* to describe research projects that "contrast among different macro-level units, such as world regions, countries, subnational regions, social milieus, language areas and cultural thickenings, at one point or more points in time" (Esser and Vliegenthart 2017, 2). The field has engaged intercultural communication perspectives since the 1950s and cross-national approaches since the 1970s, driven by an interest in cross-cultural and political communication topics (Blumler and Gurevitch 1975; Hall 1976; Hofstede 1983; Blumler, McLeod, and Rosengren 1992; Hallin and Mancini 2004; Norris 2009).[3]

Cross-national and regional scholarship in communication and media studies has been fueled by at least two interconnected insights, both emphasizing the limitations involved in single-country approaches. The first insight is the centrality of media and communication in processes of globalization (Livingstone 2012). This sparked, on the one hand, an interest in the question of cooperation across diverse national settings, as illustrated by work in inter-cultural communication (Kim 2012), and, on the other, efforts against "methodological nationalism" (Beck 2000) aiming to display a global outlook. As Sonia Livingstone argued (2012), cross-national studies showed that "it is no longer plausible to study one phenomenon in one country without asking, at a minimum, whether it is common across the globe or distinctive to that country or part of the world" (417). The second insight has been the field's recognition of a parochial and universalizing Western bias whereby "the same few countries keep recurring as if they are a stand-in for the rest of the world" (Curran and Park 2000, 2). This bias has been paired up with an uncritical uptake of cultural globalization (Curran and Park 2000; Morris and Waisbord 2001; Kraidy 2009), which has often ended up erasing the place of the state. The cross-national and regional variety of comparative research thus denaturalizes the single-country strategy and prevents, at least a priori, any nation or region to acquire a default status.

Different typologies have been put forward to assist in the process of cross-national and regional comparisons. A prominent one was developed by Geert Hofstede, who contended that "the comparison of cultures presupposes that there is something to be compared; that each culture is not so unique that any parallel with another culture is meaningless" (1984, 32). He proposed the existence of four dimensions according to which all cultures could be compared: individualism versus collectivism, uncertainty avoidance, power distance, and masculinity versus femininity (Hofstede 1983). In another highly influential cross-national approach, Daniel C. Hallin and Paolo Mancini (2004) put forward a different typology that sought to correlate nations with media systems. Analyzing eighteen Western democracies, they came up with the "Mediterranean or Polarized Pluralist, North/Central European or Democratic Corporatist, and the North

Atlantic or Liberal models" (2004, 2). This account opened the door for a series of studies that made visible a variety of media systems around the world (Sparks 2008; Brüggemann et al. 2014; Guerrero and Márquez-Ramírez 2014). In the authors' words: "As Bendix (1963: 537) says, comparative analysis has the capacity to "increase the 'visibility' of one structure by contrasting it with another." Analysts deeply steeped in one media system will often miss important characteristics of their own system, characteristics that are too familiar to stand out to them against the background. . . . Comparative analysis is essential if we want to move beyond these limitations" (Hallin and Mancini 2004, 302).

In addition to these cross-national and regional varieties of comparative research, our cross-media work has been informed by scholarship produced between the end of the twentieth century and the start of the twenty-first century aiming to conceptualize what might be distinct about what was then often referred to as "new media" (Williams, Rice, and Rogers 1988; Rice 1999; Manovich 2002; Chun, Fisher, and Keenan 2004). A recurrent theme across the different perspectives adopted to address this matter, and the resulting answers about their distinct character, was the centrality of the connections between the then new media and their older predecessors. Thus, in their influential treatise of how new media always remediate, Jay David Bolter and Richard Grusin argued that "what is new about new media comes from the particular ways in which they refashion older media and the ways in which older media refashion themselves to answer the challenges of new media" (1999, 15).

Maintaining a cultural focus on aesthetics while adding accounts of industries and audiences, Henry Jenkins drew inspiration from the pioneering assertion of Ithiel de Sola Pool that a "convergence of modes [was] blurring the lines between media" (de Sola Pool 1983, 23) to highlight the centrality of this convergence as a defining feature of the new media that entailed "the flow of content across multiple media platforms, the cooperation between multiple media industries, and the migratory behavior of media audiences who will go almost anywhere in search of the kinds of entertainment experiences they want" (Jenkins 2006, 2). Furthermore, combining social and behavioral science approaches, Leah Lievrouw and Sonia Livingstone (2002, 8) emphasized in their introduction to the first edition of *The Handbook of New Media* "the essentially continuous nature of new media development. Even technologies that are perceived as being unprecedented are found upon closer analysis to have been designed, built and implemented around existing technologies and practices." Thus, capturing the spirit of these and other related ideas, David Thorburn and Henry Jenkins (2003) criticized what they called "medium-specific approaches" and made a strong plea for a "comparative approach": "Medium-specific approaches risk simplifying technological change to a zero-sum game in which one medium gains at the expense of its rivals. A less reductive, *comparative approach* would recognize the complex synergies that always prevail among media systems, particularly during periods shaped by the birth of a new medium of expression" (3; emphasis added).

These conceptual developments relate to another trend in scholarship that blossomed during this period and that has also informed our cross-media comparative perspective: historical accounts of specific innovations in media, information, and communication technologies that inquired into dynamics of both continuity and discontinuity with previous artifacts (Boczkowski 2004; Fischer 1992; Sterne 2003; Thompson 2002; Turner 2006). In *When Old Technologies Were New*, Carolyn Marvin (1988) articulated the historiographic foundation of this kind of scholarship as follows: "New media, broadly understood to include the use of new communications technology for old or new purposes, new ways of using old technologies, and, in principle, all other possibilities for the exchange of social meaning, are always introduced into a pattern of tension created by the coexistence of old and new, which is far richer than any single medium that becomes a focus of interest because it is novel" (8).

This coexistence of old and new within broader and ongoing social, cultural, and institutional patterns became a generative lens with which to probe specific issues about distinct technological innovations. One example of particular significance to social media is Susan Douglas's (1989) account of the role of amateurs in the early development of radio. Much like the case of contemporary platforms such as Facebook, built largely by technically savvy innovators initially outside of the corporate landscape and premised upon notions of an active user base, Douglas sheds light on the key role that amateurs played in switching radio from telegraphy into a broadcasting model. Their participation amounted to "a revolutionary social phenomenon. . . . A large radio audience was taking shape whose attitude and involvement were unlike those of other, traditionally passive, audiences. . . . This was an active, committed, and participatory audience" (205). Thus, in the span of a few years, "the amateurs and their converts had constructed the beginnings of a broadcasting network and audience" (302).

Shifting from radio to print, Adrian Johns's (1998) analysis of the social relationships and conventions that undergirded the credibility of knowledge in the early days of book publishing has an uncanny significance for digital media—including platforms—especially considering issues of distrust about them. To Johns, the "political and moral economies of publishing and reading are enormously different now from their state in Newton's day. Nevertheless, a close examination will almost certainly reveal not an elimination . . . but a transformation of the kinds of sociability and civility involved" (1998, 636). For instance, he notes that "Financial institutions and other corporations are laboring to establish a means of rendering electronic communication secure enough to supplant more traditional media. It is not too fanciful to *compare* these efforts to the Royal Society's endeavors to secure the credit of printed communications in the seventeenth century" (Johns 1998, 637; emphasis added).

Supplementing these cross-national and cross-media approaches, our cross-platform work has been informed by a series of theoretical developments that since the 1980s have tried to explain how users make sense of an increasingly diversified media environment. Central to this development was the growing multiplicity of channels available on cable television, which sparked questions around issues of choice and awareness of alternatives (Webster and Wakshlag 1983; Perse 1990). Scholars articulated the notion of repertoires to address this multiplicity (Ferguson 1992; Ferguson and Perse 1993). First defined by Carrie Heeter (1985) in relation to "the set of channels watched regularly by an individual or household" (133), the notion reflected the idea that users were drawn to a relatively smaller set of options from which they made consumption decisions. The "high-choice media environment" (Prior 2005) became manageable in everyday life through the construction of media repertoires (Webster 2011). The concept has been recently applied to newer media, with studies (Kim 2016; Lin 2019) showing how repertoires are "essentially structures that are recursively activated within their daily social practices" (Taneja et al. 2012, 964).

Another influential approach developed to analyze how users deal with an ever-expanding array of media alternatives is the notion of polymedia (Madianou and Miller 2013; Renninger 2015; Madianou 2016). Initially applied to account for family communication dynamics that take place within the context of transnational migration processes, Mirca Madianou and Daniel Miller (2012) coined the term to describe "a new communication environment" (1) in which different options coexist—among them, social media— and whose use shapes, and is shaped by, family relationships. In the authors' words: "Polymedia shifts the attention from the individual technical propensities of any particular medium to an acknowledgement that most people use a constellation of different media as an integrated environment in which each medium finds its niche in relation to the others" (Madianou and Miller 2012, 3).

The notion of constellations, mentioned by Madianou and Miller, is also key to understanding how social media repertoires are enacted in everyday life. Pablo Boczkowski, Mora Matassi, and Eugenia Mitchelstein (2018) show how the constellation of meanings that users attribute to different social media platforms shapes their repertoires of practices. These meanings are relational since the meaning attributed to a given platform is partly determined by the meanings attributed to the other platforms that constitute the social media repertoire. The creation of users' "social media ecosystems" (Zhao, Lampe, and Ellison 2016; DeVito, Walker, and Birnholtz 2018) also illustrates the extent to which platforms are deeply situated within a dense and broad web of social media use in everyday life and can be associated with the longer tradition of media ecology (Innis 1964; McLuhan [1964] 2003; Ong 1982; Postman 1986; Lehman-Wilzig and Cohen-Avigdor 2004; Strate 2004; Scolari 2013) and its double approach to consider both how media can be "an environment that surrounds the subjects and models their cognitive and perceptual system" and "the interactions between media, as if they were species of an ecosystem" (Scolari 2012, 209–210).

In sum, the varieties of comparative scholarship that we have included in this section differ empirically, methodologically, and theoretically. Yet, there is a common denominator that cuts across this diversity: the centrality of comparing as the key epistemic operation that guides the inquiry. Thus, in this book we will blur the boundaries between these varieties—albeit without erasing their differences—to develop multifaceted comparative perspectives that can contribute toward de-westernized, global, cross-media, and multiplatform scholarship on social media.

Outline of the Book

The remainder of this book consists of two multichapter parts and a concluding chapter. Each one of chapters 2 through 6 opens with contrasting vignettes that illustrate the most salient aspects of each chapter's topics with highly visible events that unfolded within— and in many cases also across—several different parts of the world. Taken together, these vignettes that range from the mundane to the extraordinary, and from the relatively insignificant to the highly consequential, show some of the main ways in which social media have been appropriated in almost every sphere of everyday life, from activism to entertainment, from religion to politics, from news to gaming, and from design to regulation, among others. The decision to start these chapters in this fashion is both to signal the reticular and multifaceted imbrication of social media in contemporary society and to underscore the pragmatic currency of our comparative perspectives.

In Foundations, the first part, we will establish the empirical, methodological, and theoretical bases that emerge from an account of the existent scholarship on social media that has compared phenomena either across nations and regions, media, or platforms. More precisely, chapter 2 will focus on cross-national and regional comparisons, chapter 3 on cross-media accounts, and chapter 4 on cross-platform examinations. In each of these chapters we will introduce a selection of eight studies to show the descriptive, explanatory, and interpretive gains that accrue from adopting at least one of these three forms of comparison, even if (as has usually been the case) this has not been done as part of an explicit comparative research agenda. Thus, we will argue that a comprehensive account of this scholarship creates the foundations on which to build an agenda for comparative social media studies.

The selected twenty-four papers that we will showcase in chapters 2 to 4 aim to maximize breadth and depth in the portrayal of the existent research. In that sense, our goal is not to furnish a representative or exhaustive depiction of the field. In other words, the following pages will not present the results of meta-analyses or systematic reviews. Instead, we chose to feature studies that examined relevant social media processes in a wide array of countries and regions of the world, connecting different traditional and social media, and a multiplicity of platforms. This is a deliberate strategy to decenter a domain of inquiry that, as explained earlier, has tended to intellectually prioritize locations in Global North settings, traditional media such as newspapers and television, and platforms such as Facebook and Twitter. While helpful in the production of certain kinds of knowledge, taken together these location, media, and platform choices have unnecessarily restricted what we know about objects of study that have global reach, connect variously to a plethora of traditional media, and are embodied in a growing spectrum of platforms.

In each of the chapters we will organize the presentation of the selected research studies into four categories of analysis that allude to key elements of the research enterprise: topics, the main issues examined in each of the papers; approaches, the dominant ways of comparing social media, either implicitly or explicitly; methods, the main methodological strategies utilized; and interpretations, the typical frames used to make sense of the findings and their implications.

Establishing these empirical, methodological, and theoretical foundations enables us to probe three central concepts for understanding social media: the nation-state, traditional media, and platforms. Thus, in chapter 2 we will argue that whereas much research and commentary has highlighted the decline of national borders regarding digital dynamics in a context of globalization, it is still worth attending to the heuristic power of the nation-state. However, its explanatory role can no longer be taken for granted within the scholarly inquiry and instead should be justified as part of the process of this inquiry. Moreover, in chapter 3 we will contend that despite the typical focus on what is new about social media, our comparative perspective emphasizes the continued relevance of their traditional media counterparts in both determining the genealogy of whatever novelty there is and the coexistence of this novelty with long-standing patterns of communication artifacts, practices, and norms. Finally, in chapter 4 we will claim that cross-platform perspectives are better suited than their single-platform counterparts to counter dystopic narratives that have recently dominated accounts of social media. This is because whereas the latter are more prone to attributing strong deterministic power to technology over the agency of users, the former create more opportunities for the emergence of variance in the findings which, in turn, make more visible the interplay between the structure of technology and the agency of users.

In Pathways, the second part, we will build on the bases established in Foundations to further articulate the contours of a programmatic agenda integrating cross-national and regional, cross-media and cross-platform dimensions of social media dynamics. We will do this by focusing on two areas of inquiry that have long been central to the study of media and communication, and to the constitution of our sense of self and social relationships more broadly: histories in chapter 5 and languages in chapter 6. Aware that prior social media scholarship has on occasion delved into either area, in each of these chapters we will first acknowledge lessons learned from these antecedents and then proceed to articulate concrete epistemic visions for comparative perspectives.

Following up on the larger conceptual issues tackled in the chapters within the first part, those in the second part will address two cross-cutting intellectual tendencies that have marked the study of social media. More precisely, chapter 5 will probe the role of histories to offset the overwhelming present-day focus that has tended to dismiss the past and naturalize the present in the relevant literature. We will argue that foregrounding historical matters can help illuminate continuities and discontinuities that are fundamental to a better understanding of what might be unique about specific platforms and social media in general. This applies to both their development and their current instantiations. Furthermore, chapter 6 will address matters of language to broaden the dominant attention to English and writing in most of the scholarship. To this end, we will articulate approaches that challenge these English-language and written-text biases through an exploration of dynamics pertaining to multiple languages and to the role of the novel visual signifiers such as emoji, hashtags, and reaction buttons that have rapidly become part of the vernacular of social media and digital culture more generally. We will contend that this aids not only in bringing new languages and signifiers into view—languages and signifiers which are the norm and not the exception in everyday social media practice—but also in properly accounting for factors that might affect variance in the case of communication only in English and/or textual form.

In chapter 7 we will bring this book to a close by taking stock of lessons learned from the previous chapters and reflecting on their broader implications for scholarship on social media. A review of the main arguments presented in chapters 2 through 6 suggests the presence of a significant level of heterogeneity cutting across social media as both objects of study and the ways in which the inquiries about them have unfolded. Building upon the notion of the heterogeneity of social media, we will probe the challenges and potential of fostering comparative work that integrates two or more of the dimensions treated separately in the previous chapters: comparisons across nations and regions, across media, and across platforms. We will argue that the challenges that might hinder the potential of these various integrations stem from long-standing trends toward the intellectual fragmentation of the different traditions of inquiry that subtend the comparative analyses undertaken in each dimension. However, we will propose that by virtue of sharing the organizing principle that to know is to compare, the perspectives advocated in this book can blur boundaries between disparate traditions of inquiry and also create trading zones (Galison 1997) among them. These trading zones can lessen the downsides of intellectual fragmentation by facilitating the exchange of ideas across often unconnected domains of inquiry in ways that do not flatten their diversity.

I Foundations

2

Cross-National and Regional Comparisons

Introduction

A Feminist Anthem, from Valparaíso to Tokyo

The date is November 20, 2019, and the location is the Aníbal Pinto Square in Valparaíso, Chile—a colorful fishermen's town lately turned into a touristic destination on the Pacific Ocean. The country is in the midst of social unrest; hundreds of photographs and videos of demonstrations are taken and shared daily on social media and messaging apps. Accompanied by the sound of a bass drum and an electronic harmony emerging from a loudspeaker, a group of around fifty people takes to the street, cuts the traffic, and sings in unison "A rapist in your path." It is an intervention and performance against *machista* violence created by the feminist performance collective LASTESIS, composed of Lea Cáceres, Paula Cometa, Sibila Sotomayor, and Daffne Valdés, and based on a proposal by the collective *Fuego: Acciones en Cemento*. The performance has been inspired in a text written by Rita Segato[1]; it denounces rape culture as a political-institutional problem, and it is directed at the Chilean police force. Activists cover their eyes with black cloth bands and actively move their bodies following a choreography while singing lyrics such as "and it's not my fault, not where

I was, not how I dressed. The rapist is you." The Aníbal Pinto Square is momentarily paralyzed. Passersby stop and record the performance with their cell phones. Within seconds their videos begin to feverishly circulate in the "digital environment" (Boczkowski and Mitchelstein 2021).

Only five days later, on the "Day Against Violence against Women in Chile," the song is played by 2,000 demonstrators gathered in Santiago, the nation's capital. The video of this performance goes viral. On Facebook, WhatsApp, YouTube, Twitter, Instagram, and TikTok, users read, like, comment, share, and retweet different recordings of the performance. From Valparaíso, or "the end of the world," as Daffne Valdés calls it,[2] "A rapist in your path" becomes a global anthem that crosses borders and languages. In less than three months, the performance is reappropriated in public spaces scattered across at least fifty-two countries, from Australia to Kenya, and from Japan to the United States. It is also translated into approximately fifteen languages, including Arabic, Basque, German, Hindi, and Mapuche.[3] An interactive map created by the nonprofit organization *GEOChicas* shows the hundreds of locations around the world where it has taken place. The evidence used by this map consists of the social media posts shared by users from their own accounts, in different languages, with hashtags converging around the same issue.[4] Less than a year after the original performance, *Time* magazine names LASTESIS one of the 100 most influential personalities of the year.[5]

A Call against Systemic Racism, from Rio de Janeiro to Minneapolis and Back

Thirteen shots are heard in the middle of the night in the neighborhood of Estácio, Rio de Janeiro, Brazil, on March 14, 2018. They are fired by two individuals from one car to another. In the second car are Councilwoman Marielle Franco, her driver, Anderson Pedro Gomes, and her press agent. They have just left a political discussion event titled "Black youths mobilizing structures." The shots kill

Franco and Gomes.[6] Franco, a thirty-eight-year-old human rights Black activist, sociologist, leftist representative in the Maré region, and feminist leader of Black, indigenous, LGBTQ, and marginalized communities in Brazil and Latin America, has been violently silenced. Various human rights organizations, including Amnesty International, begin to demand justice. News media coverage and social media commentary quickly zero in on Franco's most recent tweet. In it she denounced, just one day before the crime, the responsibility of the parapolice militias in deaths occurring in the *favelas* of Rio de Janeiro, where Franco was born and grew up.[7] The tweet suggests, for those who demand justice, the potential involvement of the police and military forces in the execution of Marielle's murder.

The reaction on social media is almost instantaneous in Brazil and also across Latin America. In a post-Arab-Spring context marked by the rise of "hashtag activism" (Papacharissi and Oliveira 2012; Costanza-Chock 2014; Hopke 2015; Jenkins, Ford, and Green 2013; Tufekci 2018; Jackson, Bailey, and Foucault Welles 2020), the hashtags #MariellePresente, #MarielleVive, and #QuemMatouMarielle blend with street demonstrations organized via Twitter, Facebook, and WhatsApp.[8] People use filters on their social media profile pictures reclaiming justice for Marielle. One year later, two military police officers are arrested for the crime. But 1,000 days after the killing, it is still not known who gave the orders to proceed with the shooting in the first place. The call for justice is coordinated, once again, on social media, via the hashtag #1000DiasSemRespostas. The action on the streets now consists of placing 550 clocks with Marielle's image; their alarms ring jointly, in front of the Rio City Council, to demand an end to impunity.[9]

Almost two years and two months later, in the northern hemisphere of the Americas, another event of horror and police brutality takes place. George Floyd is a forty-six-year-old Black hip-hop musician who lives in the city of Minneapolis, in the United States. On the afternoon of May 25, 2020, a merchant accuses him of trying to pay with a forged $20 bill. The accusation is followed by a violent arrest by local police officers. For eight minutes and forty-six seconds, a white cop presses Floyd's neck against the street until he stops breathing.[10] The cameras of horrified passersby record the moment, which rapidly goes viral. The call to end police violence and the struggle for justice against the structural racism of US law enforcement, and society more generally, travels the world at lightning speed. Within a few days, Black Lives Matter marches that take place in Minneapolis are replicated variously from China to Germany, and from Iran to South Africa, among other countries.[11] Set against the background of the challenges brought by the worst public health crisis of the past 100 years, demonstrators take to the streets and the screens with unparalleled strength. Hashtags multiply and help demands coalesce: #EndPoliceBrutality, #EnoughisEnough, #Mobilize. Grassroots movements such as #FreedomFightersDC are organized on social media.[12] On June 2, 2020, there is a dispute over the so-called Blackout Tuesday, in which people are encouraged to post a black photo on their platforms to speak out against police violence and brutality toward Black lives. The conflict arises since activists argue that the circulation of black screens can hinder the usefulness of a hashtag (#BlackLivesMatter or #BLM)[13] used by those on the streets to protect themselves from potential attackers and to eschew raids by law enforcement.

The assassination of Marielle Franco has been linked to a political crime of the repressive apparatus of the Brazilian state. The killing of George Floyd is part of a long series of racist crimes perpetrated against Black bodies in the United States by the forces of law and order. Beyond their particularities and their dissimilar geographical origins, both events share common roots of police brutality and structural racism. However, the social media aftermath of these events had divergent trajectories. While the repercussions of Franco's case spread within Brazil and, to a lesser extent, Latin America, those of Floyd's case diffused more broadly across multiple continents (Shahin, Nakahara, and Sánchez 2021).

Why Comparing across Nations and Regions Matters

Social media platforms have a fundamental spatial dimension in at least two ways. First, it is possible to conceive of them as spaces in themselves: "virtual geographies," as Zizi Papacharissi (2009) labels them, with their architectures, designs, trajectories, and borders. They are places, in the theorization of Daniel Miller and colleagues (2016), that we inhabit, from which we enter and leave, and where we build our selves, interact with others, and learn about the world. Second, the geographic and material spaces in which platforms are invented, programmed, circulated, and appropriated end up shaping their design and use. A data visualization produced by the company Visual Capitalist in 2020[14] represents the usage statistics of the platforms belonging to Facebook Inc. (United States) as taking part of the same constellation of planets—far from the constellation created by Tencent (China), and by Telegram FZ LLC (Russia). Although not as distant as constellations, cultures associated with nation-states, dependent territories, or specific regions of the world have considerable weight in the ways platforms are designed, regulated, and used, as well as in the social, cultural, and political consequences of their appropriation.

Cross-national and regional comparisons are critical to illuminate the role of these spaces and to understand the similarities and differences present in their construction and adoption. This type of comparative lens shows us that a particular use of social media, such as sharing a protest song that denounces *machista* violence in a corner of Valparaíso, can have global reach and travel across different cultures and platforms. In addition, cross-national and regional comparisons highlight how certain forms of spontaneous and organized activism can diverge in their geographic spread and uptake. While the performance of LASTESIS reflects a phenomenon of convergence in the diffusion and reappropriation of the same social demand in different countries of the world, the cases of Marielle Franco and George Floyd show a divergence in the circulation of two different, albeit related, claims. This divergence was partly patterned alongside prior differences between Global South and North.

These vignettes begin to show the descriptive, explanatory, and interpretive value of accounting for commonalities and differences; continuities and discontinuities; circulation and recirculation; and local, global, and glocal uptake of social media. These comparative analyses shed light on macro-level issues such as the use of hashtags that cross borders and make visible a claim that does not concern only one country; meso-level issues such as the organization of collective action; and micro-level issues such as the use of a hashtag to protect oneself in the context of mass mobilizations against police violence and structural racism.

This chapter proceeds as follows. We will next draw upon the findings from eight studies about social media conducted in different parts of the world to illustrate the descriptive fit and heuristic power of a comparative lens focusing on dynamics across nations and regions. We will organize them in relation to the four basic categories of scholarly practice that we first introduced in chapter 1: topics, approaches, methods, and interpretations. After making sense of some salient threads across these four categories, we will bring the chapter to a close by reflecting on the continued worth of the concept of the nation-state to make sense of platforms, an object of inquiry that crosses borders with an ease and force like no other media before.

Topics

Scholarship about a wide array of topics within communication and media studies has produced cross-national and regional comparative accounts of social media practices (Chu and Choi 2010; Jackson and

Wang 2013; Qiu, Lin, and Leung 2013; LaRose et al. 2014; Nielsen and Schrøder 2014; Miller et al. 2016). Two recurrent topics of interest have been ideological polarization and political debate—both of them critical to social deliberation in the contemporary polity.

One of the common concerns associated with social media has to do with "filter bubbles" and "echo chambers" (Sunstein 2009; Pariser 2011; Colleoni, Rozza, and Arvidsson 2014; Flaxman, Goel, and Rao 2016; Dubois and Blank 2018; Bruns 2019). These notions point to the idea that because platforms allow us to choose our audiences, and their algorithms presumably favor homophily (McPherson, Smith-Lovin, and Cook 2001), our ability to confront ideas that are inconsistent with our worldviews could tend to diminish over time (Slater 2007; Himelboim, McCreery, and Smith 2013; del Vicario et al. 2016; Entman and Usher 2018; Ling 2020; Parisi and Comunello 2020).

Marko M. Skoric, Qinfeng Zhu, and Jih-Hsuan Tammy Lin (2018) address this matter by inquiring into the dynamics whereby a person decides to stop either being friends with or unfollow another person on social media as a product of ideological disagreement. This phenomenon is known as "political unfriending" (John and Dvir-Gvirsman 2015; Bode 2016; Yang, Barnidge, and Rojas 2017; Bozdag 2020; Trevisan 2020) and, more broadly, as "selective avoidance" (Liao and Fu 2013; Messing and Westwood 2014; Bakshy, Messing, and Adamic 2015; Zhu, Skoric, and Shen 2017; Vraga and Tulli 2019). To this end, the authors compare key significant variables behind the motivations of political unfriending on Facebook and Twitter in Taiwan and Hong Kong, which they characterize as two Asian societies with common roots but also dissimilar political and cultural traits. Skoric and colleagues speculate that "as unfriending and unfollowing on social media resembles and signals the dissolution of social ties, it may be governed by cultural norms" (2018, 1,102). More precisely, that "users who endorse collectivistic values may be less likely to unfriend others in order to maintain harmony in the network" (Skoric, Zhu, and Lin 2018, 1,103).

The theoretical framework utilized by these authors includes ideas, introduced in chapter 1, about the differences between individualistic and collectivistic societies originally proposed by Geert Hofstede (1983, 1991, 1998). From this perspective, Taiwan is categorized as a "highly collectivist" society (Skoric, Zhu, and Lin 2018, 1,103) that avoids uncertainty and shows affinity with institutional hierarchies. Despite being characterized also as collectivist and hierarchical, Hong Kong's culture avoids uncertainty to a lesser degree, which to the authors indicates a greater capacity to deal with ambiguity and to be flexible when interpreting rules. To address political unfriending in these two cultures, authors analyze results from an online survey conducted in 2016.

Skoric and colleagues find that there are only small significant differences in the level at which the phenomenon of political unfriending occurs in both societies; however, they do find larger differences when it comes to the factors motivating political unfriending in the first place. In the two cases, they observe an inverse association between degree of collectivism and chances of political unfriending on social media. In the authors' words, "this is in line with the literature on collectivism, which argues that individuals strive to achieve group harmony rather than satisfy their own needs" (Skoric, Zhu, and Lin 2018, 1,110). Skoric and colleagues also note that in Taiwan—a nation with comparatively higher levels of democratic participation and social peace and in which the use of platforms is highly associational—psychological or social factors, such as the tendency toward FOMO (fear of missing out), predominate when it comes to breaking a political bond. Instead, in Hong Kong—a country with a relatively higher degree of political conflict where social media use is relatively more tied to engagement in politics—political interest has the greatest impact on the decision to stop being a friend or follower of another user. Thus, "political unfriending and unfollowing in Hong Kong are indicative of political tribalism and a symptom of heightened affective polarization present in the current Hong Kong society" (Skoric, Zhu, and Lin 2018, 1,110).

The second illustration relates to the rise of platforms in the first decade of the twenty-first century, which was coupled with a utopian perspective that imagined these spaces as conducive to a Habermasian deliberation of ideas, democratic rights, and collective action in a way not mediated by traditional political spaces or figures such as parties, unions, and opinion leaders (Papacharissi 2010; Bennett and Segerberg 2012; Halpern and Gibbs 2013; Jenkins et al. 2016). Over time, this utopian vision, partly inspired by the "rhetoric of the technological sublime" (Marx 1964), was challenged by perspectives that tended to attribute negative consequences to platforms, linking them, for instance, to the breakdown of social ties resulting from the homophily of algorithmic design (Sunstein 2017; Vaidhyanathan 2018). However, questions about the ability of platforms to promote public deliberation and participation remain open in social media studies.

Tanja Estella Bosch, Mare Admire, and Meli Ncube (2020) examine the use of Facebook for political discussion in Zimbabwe and Kenya. Both countries represent cases that "have endured decades of authoritarian rule" (352) as well as been "at the forefront of appropriating digital media platforms for political activism and campaigns" (353), especially in the face of traditional media censorship. According to the authors, the traditional spaces for debate in Zimbabwe and Kenya tend to be closed to youth, which in turn generally exhibit low levels of political efficacy and attitudes of apathy toward their political systems. The authors analyze the Facebook pages of two politicians, Emmerson Mnangagwa and Uhuru Kenyatta.

Bosch and colleagues find that in both countries Facebook pages allow citizens to engage with content shared by politicians, initiate debates around it, and perceive the possibility of participating in the extended public sphere. For example, when it comes to Zimbabwe, they observe that "citizens' concerns are being shared, heard and debated on the Internet and social media, which is making it possible to distribute and receive alternative sources of information to government propaganda, disinformation and secrecy" (Bosch, Admire, and Ncube 2020, 359). However, they also find that this does not necessarily imply a change of position with respect to presidents' mandates in their communication with the electorate. In the authors' words, "if ever there is anything to note, there is 'passive listening' whereby politicians use these invited and invented spaces of participation to monitor, predict and observe public opinion formation" (Bosch, Admire, and Ncube 2020, 361). Whereas in the case of Zimbabwe, Facebook's relevance for the public sphere increases in the face of political limits to freedom of expression, in Kenya, "the lack of dialogue between citizens and the presidency . . . represents a missed opportunity to engage in dialogue with citizens" (Bosch, Admire, and Ncube 2020, 360).

Focusing on ideological polarization and political debate, these two studies tackled central theoretical ideas in scholarship about social media such as echo chambers and engagement in the public sphere. In both cases, the main findings emerged because of the comparative perspective. Without this approach, Skoric, Zhu, and Lin (2018) would not have been able to identify the strong cultural aspect to political unfriending, and Bosch, Admire, and Ncube (2020) could have missed the intersection between national contexts, Facebook affordances, and their uptake in fostering public debates.

Approaches

We identify two central approaches in cross-national comparative studies, which can be characterized in institutionalist and culturalist terms. The first has to do with comparing the political systems of the countries or cases analyzed (Gainous, Wagner, and Abbott 2015;

Kalogeropoulos et al. 2017; Saldaña, McGregor, and Gil de Zúñiga 2015; Chen, Chan, and Lee 2016; Mosca and Quaranta 2016; Boulianne 2020). The second is based on contrasting national cultures (Chu and Choi 2010; Kim, Sohn, and Choi 2011; Jackson and Wang 2013; Katz and Crocker 2015; Trepte et al. 2017; de Lenne et al. 2020).

Regarding the first approach, political systems are conceived of as independent variables that then affect media systems—operationalized as the dependent variables—in their respective nations. A widely circulated example we mentioned in the first chapter is *Comparing Media Systems: Three Models of Media and Politics* (Hallin and Mancini 2004). The underlying logic that there is some significant relationship between a country's political system and the way its media system behaves also permeates scholarship on social media. This is even after taking into account the limitations that a national system might have to influence a sociotechnical infrastructure that has largely emerged in the Global North, has planetary reach, and can potentially be used by anyone largely regardless of the location from which they do so.

Does national political culture affect how a leading national newspaper's newsroom adopts and uses Facebook or Twitter? Jeslyn Lemke and Endalk Chala (2016) compared news media's uptake of social media platforms in Senegal and Ethiopia, two countries with variations in at least four aspects relevant to the topic under analysis: their level of democratic quality, the dominant language used by the news media, the degree of internet adoption, and their internet governance policies. Lemke and Chala argue that "we assumed that differences or similarities in social media feeds can be intangibly connected to Ethiopia's restrictive laws or Senegal's democracy" (2016, 182).

The authors examine the content produced by the Facebook and Twitter accounts of five newspapers in each country for sixty consecutive days in 2015. These posts are analyzed according to two variables: number and format. In theoretical terms, the paper is based on Mark Deuze's claim (2003) that the three central characteristics that distinguish online from traditional journalism are multimediality, interactivity, and hypertextuality. Overall, Lemke and Chala find a mix of commonalities and differences between the two countries attributed to their respective political systems. On the one hand, both countries share the way journalists use platforms for storytelling purposes. Far from Deuze's goal of multimedia journalism, they find that content shared on social media platforms tends to replicate the information produced for the print or digital version of newspapers: "In the ten newspapers we analyzed in Senegal and Ethiopia, 'networked' journalism seems to be on the horizon, but creating specialized content for a newspaper's multimedia platforms is yet to come" (Lemke and Chala 2016, 180–181). On the other hand, although in Senegal, platforms are used to channel the same content into different traditional and social media, in Ethiopia, Facebook is mostly used for international news. The authors hypothesize that this difference may be because Ethiopian newspapers strategically use this platform to include foreign media links and thus increase freedom of expression without running the risk of political persecution.

The second approach commonly present in cross-national studies of social media has a cultural bent. Following Stuart Hall (1980), culture can be understood as "the categories and frameworks in thought and language through which different societies classif[y] out their conditions of existence" (65). Venetia Papa and Dimitra L. Milioni (2016) compare issues of national culture in Facebook groups based in Greece and France regarding the *Indignados* movement. This movement emerged spontaneously around 2011 in different countries to fight against political and economic corruption and to claim for the rights of the unemployed and the disenfranchised (Castañeda 2012; Anduiza, Cristancho, and Sabucedo 2014; Postill 2014; Flesher Fominaya 2015; Theocharis et al. 2015). According to Papa and Milioni (2016), the movement represents a compelling case study because it appeals to an international collective that, while not responding to a clear ideology, becomes visible against the backdrop of national differences. In addition, *Indignados* is particularly relevant for social media scholarship since social media platforms became for this movement a critical space for self-organizing and increasing public visibility. The authors concentrate on Greece and France because each country shows different roles in the evolution of this social movement. In Greece, it involved a series of anti-austerity protests against the tightening of financial policies and living conditions for workers,[15] which were in turn supported in France a few months later. From France also came the political pamphlet turned into the book *Time for Outrage: Indignez-vous!* (Hessel 2011), which contains ideas believed to have inspired *Indignados* in Spain, a country central to the development of the movement.

Based on a thematic analysis of content on Facebook postings and in-depth interviews with activists, Papa and Milioni (2016) find that the way activists recognize and relate to each other online has to do with a notion of citizenship that exceeds geography and ideology: "[A]s the *Indignados* movement is void of a predefined political identity, a certain (defiant) understanding of civic identity becomes the motive or the 'social glue' that brings them together" (296). They also find that the central trait of the movement, that of including the excluded, reinforces the non-national condition of movement membership on social media. However, in the case of Greek activists, Papa and Milioni (2016) note how the demand for nationalism emerges from some radicalized members and is directly linked to discourses referring to a "homogeneous Greek state" (301). Specifically, "through their discourses, they express their strong bond with an idealized Greek nation, directed by the need to 'save their nation' from internal and external enemies" (Papa and Milioni 2016, 297). Regarding French activists, authors find a rhetoric by which "individuals who are mostly citizens of Maghreb countries . . . do not possess the formal status of French citizenship" (Papa and Milioni 2016, 297).

Both cases show the different meanings that "we" can take within the same political movement.

In this section we have showcased two approaches to the varying roles played by national and transnational variables in illuminating the adoption of specific platform practices across countries. In both cases, the comparative perspective was fundamental. Lemke and Chala's interpretation of the finding of Ethiopian newspapers using Facebook for international news was possible thanks to the contrast of Senegalese newspapers' use of Facebook and the consideration of the Ethiopian political context. Papa and Milioni's capacity to observe both a virtual commonality transcending geographical borders and divergent national enactments of a single social movement was enabled by their comparison of discourses of social media users in two different cultures.

Methods

One important methodological dimension in comparative cross-national and regional research has to do with the volume and kind of data analyzed. On the one end we note large-N studies that mostly rely on surveys (Jackson and Wang 2013; Ku, Chen, and Zhang 2013; Nielsen and Schrøder 2014; Trepte et al. 2017; Skoric, Zhu, and Lin 2018). On the other end we observe accounts that draw upon small-N data (Miller et al. 2016; Papa and Milioni 2016; Kalogeropoulos and Nielsen 2018; Abokhodair and Hodges 2019; Masullo et al. 2020).

Concerning the former, Dustin Harp, Ingrid Bachmann, and Lei Guo (2012) focus on understanding "more about activists who use the Internet and social media, their perspectives on these new technologies, and the scope of their work" (299) and on the variation of these issues across three distinct locations: Latin America, mainland China, and the United States. Their ultimate interest resides in providing a comparative perspective on the ways in which digital public spheres are shaped in different regions. The authors "treat these regions as three separate cultures or systems of meaning comprised by shared beliefs, norms, and expectations" (Harp, Bachmann, and Guo 2012, 302).

Their goal is to examine research that has criticized online activism for its lack of "real life" engagement or questioned the actual inclusivity of the digital public sphere. Analyzing online surveys administered in Chinese, English, and Spanish, Harp and colleagues find significant differences in the ways in which activists conceive of social media when it comes to managing them and assessing their capacity to shape the digital public sphere: "For respondents in China, the top challenge for using SNS [social networking sites] for activism was fear of government surveillance, while for those in the United States, it was the lack of time. Respondents from Latin America, on the other hand, emphasized the lack of access to affordable Internet, and, indeed, 15% of these survey respondents said they did not have access to the Internet in their own homes" (Harp, Bachmann, and Guo 2012, 313).

Thus, Harp, Bachmann, and Guo (2012) conclude that "social media can become a participatory forum where people with common interests can come together, become empowered, and ultimately join efforts to improve their communities" (314). On the basis of these findings, they explain that their "cross-cultural approach allows for a more nuanced understanding of the phenomenon" (2012, 314).

An illustration of the small-N alternative is Cigdem Bozdag and Kevin Smets's (2017) examination of the reception of the image of a deceased Syrian boy named Alan Kurdi. Turkish photojournalist Nilüfer Demir captured Kurdi's dead body found on the shores of the Mediterranean Sea on September 2, 2015, in the midst of an ongoing refugee crisis.[16] Bozdag and Smets decide to analyze the circulation of Kurdi's image on Twitter in Flanders and Turkey resulting from posts by different social actors: citizens, politicians, and nonprofit organizations. Both settings are selected because they are either geographically close to Syria, the country most associated with the recent refugee crisis, or because they are refugee-receiving countries. Specifically, the authors examine a corpus of 961 tweets, using both inductive and deductive codes, and pay special attention to how refugees and migrants are represented in each case. Among the codes, Bozdag and Smets include whether refugees are represented in individualized or collectivized ways, the reasons around the refugee crisis, its proposed solutions, and references to the case of Alan Kurdi.

The authors find that far from producing a single, global understanding of Kurdi's image as a symbol of a humanitarian crisis, the meanings produced in each case were strongly shaped by the national context of reception. There were issues in common—such as the association of the photograph with the presentation of refugees as a threat to national order and security. However, there were also two important differences. First, in Flanders there was much more interaction across social media posts than in Turkey. The authors connect this difference with the level of social polarization in each context: "[P]ublic perceptions of immigration take shape in a broader context of societal polarization in Turkey (in relation to ethnicity, religion, and politics), whereas in Flanders, there is a rather dominant anti-migration and anti-Islam discourse, nourished by decades of polarization of the extreme right" (Bozdag and Smets 2017, 4,056). Second, religion—operationalized by the authors as either the belief or the nonbelief in Islam—shaped whether the image of Kurdi was interpreted in either a sameness key or as an example of otherness: "When reference is made to Islam in Turkey, it serves as a vehicle for solidarity and a religious obligation to help other Muslims. In Flanders, Islam is mentioned by certain politicians and citizens who explain it as the source of cultural differences" (Bozdag and Smets 2017, 4,064).

There has been a range of designs patterned along the dimension of the volume and kind of data utilized in comparative cross-national and regional work. In the two examples we showcased in this subsection the analyses revealed findings that would have probably remained invisible through single-country accounts. In comparing social media perceptions from Latin America, mainland China, and the United States via a survey, Harp and colleagues (2012) were able to observe the relative importance of, for instance, internet access for activists. In contrasting two contexts via a qualitative content analysis, Bozdag and Smets (2017) showed how they strongly shaped divergent social media representations of the refugees and the refugee crisis.

Interpretations

A popular interpretive frame in cross-national and regional comparative studies of social media is making sense of the findings in terms of either divergence or convergence of phenomena under examination. On the one hand, there is the notion that under certain circumstances the culture associated with the national territory effectively shapes the use of platforms and ends up producing significant variations. On the other hand, there is the idea that despite the differences among countries, there are major points in common in the use of social media.

According to the "protest paradigm," traditional media tend to cover news linked to social mobilizations usually with a reactionary and right-wing bias that has a detrimental impact upon the public legitimacy of protests (Gitlin 1980; Chan and Lee 1984; McLeod and Hertog 1992; Harlow and Johnson 2011; Mourão 2019). How is this paradigm applied in the context of social media? To answer this question, Summer Harlow (2019) investigates coverage of the protests in Ferguson, Missouri, United States, that took place in reaction to the murder of Michael Brown by local police on August 9, 2014. Examining data from four countries—France, Spain, the United Kingdom, and the United States—Harlow looks at how the focus on racism and police brutality shapes the framing and perception of protests. To this end, she compares the tweets produced by media organizations, journalists, and the public.

Harlow finds that across these four countries, media organizations tend to highlight the issue of police brutality. This in turn downplays the relevance of structural dynamics regarding racism in law enforcement and sidelines the core theme that organizes and legitimizes protests in the first place, since "focusing on police brutality rather than racism painted the issue as a problem specific to individual cops rather than systemic racism deserving of protest. The underlying reason for protests thus was ignored, as the protest paradigm would suggest" (Harlow 2019, 635). Beneath these commonalities Harlow notes that whereas in France and the United States journalists tend to present post-racial views, in Spain and the United Kingdom they emphasize racism as the core target of social mobilization. The author attributes this divergence to the historical memory of the latter two countries concerning racial inequality. When it comes to the protest paradigm and to how demonstrators are framed, Harlow notes that "the U.S. media outlets and their journalists' tweets adopted delegitimizing frames of protesters significantly more than the U.K., Spain, and France" (2019, 636). According to Harlow, "this finding illustrates the importance of comparative research and the need to better understand how the paradigm changes on a country-by-country basis" (2019, 636).

In what she sees as a context of transformation for digital journalism, Amy Schmitz Weiss (2015) investigates how journalists in Argentina, Brazil, Colombia, Mexico, and Peru link news production routines to social media practices—whether in the newsroom or in the context of individual coverage. To examine how "legacy and non-legacy media organizations . . . are facing dramatic changes to the news production and distribution process," the author analyzes responses to an online survey and follows how different journalistic cultures are perceived and enacted by reporters and editors.

According to Schmitz Weiss, "it is only by continuing to do comparative research that we can see how these different journalistic national cultures differ and how they are similar" (2015, 96).

Survey results indicate that how journalists appropriate social media platforms in their production routines is linked to their professional roles, which are in turn partly shaped by national contexts. Schmitz Weiss distinguishes four roles that function as Weberian ideal types (Weber 1949): adversarial, interpretive, disseminator, and populist mobilizer. The first has to do with presenting an adversarial position to political and economic interests. The second connotes that the journalist must focus on analyzing and interpreting complex phenomena. The third is conceived as a provider of information in ways that educate and entertain. The fourth espouses a normative position that sets the public agenda, informs audiences, and proposes solutions to societal problems.

The five countries studied present commonalities in terms of their political infrastructures, but they also show areas of divergence in relation to their media systems—in both media ownership and state intervention. However, Schmitz Weiss finds that, across case studies, journalists identify more with the interpretive and with the populist mobilizer roles, and considerably less with the disseminator and adversarial roles. This, in turn, shapes the digital media routines engaged in their everyday professional tasks. For instance, the populist mobilizer role was more associated with the task of searching news releases. She states: "Considering all five countries showed significance in this area [populist mobilizer] demonstrates how much the journalists surveyed in this study perceive a different role than just an informer or disseminator that can be tied back to the unique media evolution that is now taking place in each of these countries" (Schmitz Weiss 2015, 94). The author also notes that journalistic roles, which are associated with culture, change over time and with everyday practice: "[R]oles are not stagnant but may change as the journalist's work changes. . . . These roles may need to be adjusted to new ways of looking at the profession" (Schmitz Weiss 2015, 94–95).

In this section we highlighted the coexistence of divergence and convergence interpretive frameworks. On the one hand, Harlow's (2019) study showed significant differences between France, Spain, and United Kingdom, and the United States regarding how newspapers and journalists on Twitter covered the events following the murder of Michael Brown in Ferguson, Missouri. On the other hand, Schmitz Weiss (2015) found how, despite some differences, journalists from Argentina, Brazil, Colombia, Mexico, and Peru shared many commonalities in relation to their imagined professional roles. Both interpretive frameworks demonstrate the descriptive and heuristic importance of comparing cross-nationally. Neither the divergent nor the convergent dynamics could have been foregrounded without examining social media processes across countries.

Conclusions

This chapter presented comparative research on twenty-two countries—Argentina, Belgium, Brazil, Chile, China, Colombia, Ethiopia, France, Germany, Greece, Hong Kong, Kenya, Mexico, Peru, Netherlands, Senegal, Spain, Taiwan, Turkey, United Kingdom, United States, and Zimbabwe—spanning four continents: Africa, the Americas, Asia, and Europe. The studies crossed borders and addressed a multiplicity of cultural, social, political, and technological formations: from communication dynamics that privilege a sense of collective harmony to modes of sharing content on platforms that express political dissent; from quasi-dictatorial regimes to liberal democracies; and from levels of internet connectivity and access reaching below 6 percent of the population to almost universal uptake of mobile devices. In all the studies we discussed how it would have been impossible to account for variance in the phenomena under examination without resorting to comparative work. In addition, the variety of methodological and theoretical resources used across these studies underscores the idea, first introduced in chapter 1, that this work can encompass an array of choices within a broader epistemological stance.

In some cases, the countries studied were compared because they belonged to the same geographical region. This is associated with the research design of the "most similar systems" (Collier 1993, 111). Since, as Arend Lijphart (1971) explains, the comparative method runs the risk of presenting more variables than compared cases, one way to proceed is to select cases where contexts present the lowest number of differences possible; this allows the differences identified to be effectively used to attribute causality. In other occasions, the countries selected are compared precisely because they rank very differently on a specific dimension. This relates to the research design of the "most different systems" (Teune and Przeworski 1970, 34), by which "different contextual conditions . . . are used to explain different outcomes regarding the object under investigation" (Esser and Vliegenthart 2017, 3).

The notion of countries as units of analysis is a core element of the interpretive and explanatory apparatus of cross-national or regional accounts. As mentioned in chapter 1, this foundation is based on the role of communication processes in the constitution of the nation-state and in media innovations related to globalization. Benedict Anderson (1983) argued for the importance of communication in the historical emergence of the nation-state as an "imagined community." He also elaborated on the role of technological change in the joint evolution of nationalism and everyday communication, a theme which he continued to explore in subsequent writing (Anderson 1994). Communication has a constitutive relationship to the nation because the latter is seen as an imagined community that is talked about, circulated, and questioned in the interactional and mediated practices of everyday life, as Mick Billig (1995) explained in his analysis of "banal nationalism." Within the context of this chapter, the massive adoption and varied use of social media in all continents over the past decade brings up the following question: What is the validity of the nation-state as a reservoir of heuristic power for making sense of communication phenomena in a world that is increasingly deterritorialized (Appadurai 1990) and traversed by platforms of planetary reach?

The question of the validity of the nation-state is linked to an ongoing debate about technological change and globalization in traditional and digital media (Boczkowski, Mitchelstein, and Walter 2011; Schroeder 2016; Hallin and Mancini 2017; Schünemann 2020; Steinberg 2020). On the one hand, a video like the one produced by LASTESIS at the almost literal end of the world can be a tool for the replication of communication practices and social mobilizations across continents, albeit with local adaptations. On the other hand, the propagation of related claims against racist violence such as those following the murders of Marielle Franco and George Floyd can follow dissimilar trajectories that reproduce divergent cross-national and regional patterns of information flows. The future of comparative cross-national and regional research on social media lies partly in deciphering under what conditions and by which mechanisms these different dynamics take place and what implications this has for the validity of the nation-state as a source of heuristic power.

Many of the challenges encountered in comparative work at the cross-national or regional level have to do with the impact of globalization, which is fueled by media and communication processes and which seems to question the capacity of nation-states to operate as either objects of study, contexts of study, or units of analysis, an issue summarized by Sonia Livingstone (2003) as follows: "Given the tension between theories of media, culture, identity and globalization on the one hand and the crossnational interests and frameworks of research funders, policy-makers and research users on the other, any project seeking to conduct cross-national comparisons must surely argue the case for treating the nation as a unit, rather than simply presuming the legitimacy of such a research strategy" (480).

As this chapter shows, scholars continue placing the nation-state at the center of their theoretical apparatus designed to explain different social media phenomena. Therefore, the studies analyzed, echoing hundreds of other comparable studies, point to at least a tentative answer to the question posed above: The heuristic power of nation-state is still worth considering, but its validity should not be assumed and taken for granted. Instead, it should be demonstrated as a result of the research process.

The comparative perspective we have presented would be incomplete, however, if we did not refer to another kind of comparison that is important for understanding phenomena linked to social media. When users tweet about their presidents or find a news item about the political arena that leads them to stop being friends with a contact on Facebook, they do not do so only in relation to the cultures of the nation-states or regions to which they belong. They also undertake these practices in connection to the cultures, structures, and institutions tied to another central element of modern societies: traditional media. It is the relationships between traditional and social media practices that we turn to next.

3
Cross-Media Comparisons

Introduction

Fame Exceeds a Single Medium

Kylie Jenner, Kendall Jenner, and Khloé Kardashian hire professional makeup artists to transform their faces with prosthetic elements. Their goal is to go out on the streets of Los Angeles disguised as ordinary people and eventually buy a smoothie without being identified by the paparazzi. This is a scene from Season 12 of *Keeping Up with the Kardashians*, a reality show that aired on cable channel E! from 2007 until 2021. Originally conceived to present, amid the mundane and the sassy, the everyday life of a wealthy but initially not famous family, the show ended up launching each of its members into global stardom.

Keeping Up with the Kardashians is the product of an era in which the logic of traditional media was gradually beginning to coexist with that of new media (Bolter and Grusin 1999; Manovich 2002; Thorburn and Jenkins 2003; Chun, Watkins Fisher, and Keenan 2005; Jenkins 2006; Douglas and McDonnell 2019). Adding a twist to the aesthetics of the film *The Truman Show*, reality television proposed a novel format in the media ecosystem: It placed viewers at the center of the television stage, focusing on their everyday realities. As Susan J. Douglas and Andrea McDonnell (2019) argue about MTV's iconic show *The Real World*, "it reimagined for television a trope previously on display in cinema and radio—the ordinary person, plucked from obscurity, thrust into the spotlight" (230). Consistent with this innovation in television aesthetics, in 2006 *Time* published a historic cover in which it announced that it had named Person of the Year none other than the magazine's reader. It was around that time that Facebook opened its doors to any user thirteen years of age or older who had a verified email address, regardless of whether they were enrolled at a university.[1] Shortly before that, YouTube had launched with the video *Me at the zoo*. In it, an ordinary individual—in fact, one of the platform's founders—talked on camera about how cool the trunks of the elephants were at the San Diego Zoo.[2] According to the trend of being mundane that Dhiraj Murthy (2018) identifies for the inaugural messages of different communication technologies in the nineteenth and twentieth centuries, the video had no purpose other than recording the ordinary (Strangelove 2010; Marwick 2013; Arthurs, Drakopoulou, and Gandini 2018; Burgess and Green 2018). The early 2000s marked the beginning of what Paula Sibilia (2008) calls the "show of the self," reinforcing the "me, me, me culture" examined by Silvio Waisbord (2020) in his analysis of the central myths and tensions of American society.

Despite the prosthetic makeup, the paparazzi finally spot the Kardashian sisters. Before the scene concludes, Kylie takes a selfie and says something that makes visible connections and tensions across media: "I think I'm gonna Snapchat before the paparazzi sell the photo . . . They can't get the first look." In the format of the typical reality television confessional, in which the protagonists stand in front of the camera and narrate in voice-over the events in the screen, the makeup mogul explains, "we are just gonna post on social media so that *we* get the story out there first and they can't twist it into their own words."[3]

Native to the small screen, the Kardashians-Jenners have also been increasingly recognized as central figures on social media. At the time of writing this chapter they have more than 1.25 billion users on Instagram and top the lists of the most followed influencers in the world. Their uses of these platforms have been tied to significant changes in the construction and image of celebrities and micro-celebrities worldwide, from the normalization of the selfie to the application of filters, and from influencer marketing to the recording of lifestyles, in a practice that Lee Humphreys (2018) traces back to nineteenth century communication patterns in the United States. In 2020, it was announced that *Keeping Up with the Kardashians* would conclude after fourteen years and twenty seasons. However, it was quickly recognized that this would not end the careers of the sisters. A year earlier, the *New York Times* had published a related essay titled, "When Instagram Killed the Tabloid Star."[4] Another article in the same newspaper explained that Kim Kardashian accumulated more followers on her Instagram account than all the combined accounts of the *Condé Nast* media conglomerate—publisher of iconic titles of contemporary print culture such as *Vogue*, *The New Yorker*, *Vanity Fair*, *Bon Appetit*, *GQ*, and *Wired*.[5] Who needs ink on paper anymore when one can read stories on the 'Gram? As Henry Jenkins (2006) put it, "in the world of media convergence, every important story gets told, every brand gets sold, and every consumer gets courted across multiple media platforms" (3).

About Streamers and Reporters

It is 9:00 p.m. eastern time on Tuesday September 29, 2020. The first US presidential debate of that year's electoral cycle begins. Hasan Piker, a young progressive from New Jersey, is already live on his Twitch channel. He broadcasts from what appears to be the living room of his home, with a professional microphone, seated in a gamer-style chair. A portion of a Bernie Sanders poster can be identified within the cluttered background of his rectangular screen.

Piker's goal that night, as well as throughout the week, will be to stream his reactions and political commentary for an audience that will probably interact in the form of texts, emoji, and memes.

He is joined by two streamer-commentators, also from their respective homes. Most of the time, the three participants remain in silence, listening attentively to the debate between Donald Trump and Joe Biden. Unlike television anchors, who are speaking from studios across the United States and many countries around the world, Piker does not rest his eyes on the camera to look at the viewer. His gaze is pointed at his own computer, with various tabs opened on the screen, from where he monitors multiple platforms, reads news aloud and, of course, follows the debate itself—which is being streamed live on *CNN*'s YouTube channel.

Piker has been a columnist for the *Huffington Post* and a producer, host, and journalist for the YouTube show *The Young Turks*. That show originally started as a radio program, then migrated to YouTube, and eventually got airtime on television signals and streaming services. According to Wikipedia, it streams on "Amazon Prime Direct, iTunes, Hulu, Roku, on Pluto TV through a 24-hour feed and on social media platforms Instagram, Facebook, and Twitter."[6] During the 2020 US presidential election week, Piker was Twitch's most popular streamer, racking up 6.8 million hours watched.[7] A significant portion of those hours were most likely tied to the stream he conducted remotely with Democratic Congresswoman Alexandria Ocasio-Cortez to encourage voter registration in the United States. The meeting did not consist of a solemn debate on the civic responsibility of exercising the right to vote in a democratic election. Instead, it had to do with playing *Among Us*, one of the most popular multiplayer games of recent times.[8] This should not be surprising: Twitch is considered a gamer-friendly social media platform, where part of the core appeal lies in watching and interacting online with amateur and professional gamers (Taylor 2018). However, a significant part of the regular interaction on the platform has recently also turned toward politics and social activism. During the *Black Lives Matter* protests of mid-2020, Twitch became an important space for collective organizing and political activism.[9]

In 2020, the *New York Times* published a profile on Piker, contrasting the authenticity and closeness offered by figures like him on platforms such as Twitch with the more manufactured and distant personas typically associated with political presenters and commentators on traditional television.[10] Curiously, the domain of this platform, now bought by Amazon, is .tv, and its presentation resembles that of a television screen. One does not have to have an account or be logged in to scroll through Twitch's live streams—on the contrary, accessing the content is similar to turning on a television set. In the words of Andrew Chadwick (2017), "older and newer media logics in the fields of media and politics blend, overlap, intermesh, and coevolve" (5).

Engaging in Comparative Work across Traditional and Social Media

There is a common thread between the stories of the Kardashian-Jenner family escaping the paparazzi via Snapchat and Piker commenting from his living room about the US presidential debate on Twitch. From the national birthplace of global entertainment, celebrity, and showbiz culture (deCordova 1990; Gamson 1994; Glynn 2000; Murray and Ouellette 2004; Marwick 2013; McClean 2014; Douglas and McDonnell 2019), the personal becomes public in the case of the Kardashians-Jenners, and the public becomes personal in the case of Piker. In addition, the two situations not only reflect complex transmedia phenomena (Jenkins 2006; Scolari 2009; Evans 2011; Jenkins, Ford, and Green 2013), where multiple interactions take place across media and platforms, but also illustrate the heuristic power of comparative work. This is because to understand the practices undertaken around one medium or platform, it is necessary to compare them with the practices enacted in relation to other media and platforms. Comparing forms of representation and practice of the Kardashian-Jenner family and Piker across various media and platforms reveals commonalities, differences, and particularities while also illuminating processes of cross-media transformation.

The Kardashian-Jenner vignette tells a story of feedback loops between the logics of the different media involved—consistent with one of the studies later discussed in this chapter (Dubrofsky 2011), which argues that the 1990s reality television partly created the ethos of social media platforms as we know it, and in turn the platform practices triggered recent innovations in reality television. The case of political commentators like Piker foregrounds dynamics of displacement whereby social media seem to occupy a place left vacant by traditional media. Neither the feedback loops nor the displacement dynamics would be adequately intelligible without a comparative gaze, which removes social and traditional media from self-contained analyses and places them in a relational perspective.

In what follows we present eight selected studies that showcase key issues of cross-media comparative work. We will organize them in relation to the four categories stated in chapter 1: themes, approaches, methods, and interpretations. We will conclude the chapter with an analysis of the contemporary relevance of traditional media in establishing the genealogy of the new and the continuing influence of the old, even in—or perhaps because of—a networked society (Castells 2004; van Dijk 2006; Rainie and Wellman 2012; Marwick and boyd 2014).

Topics

Two recurring topics in cross-media scholarship have been the relationship between different media and the political realm and the relationship between different media and journalistic practices. This is not entirely surprising if we consider the historical link between political science and comparative theory and its strong connection with some of the first comparative studies in the field of communication (Blumler and Gurevitch 1975; Gurevitch and Blumler 1990; Blumler, McLeod, and Rosengren 1992; Norris 2009; Esser and Hanitzsch 2012; Esser 2019). Research comparing traditional and social media and politics has explored electoral campaigns, public debates, and governmental communication, among others (Benoit et al. 2011; Skoric and Poor 2013; Kalsnes, Krumsvik, and Storsul 2014; Chadwick 2017). The relevance of cross-media matters to journalism practices is also unsurprising because media organizations have been experimenting with social media for well over a decade now. Some key areas of inquiry have been the refashioning of editorial routines, the dynamics of inter-media agenda setting, and the evolving practices of news reception, among others (Neuman et al. 2014; Schrøder 2015; Abdenour 2017; Harder, Sevenans, and Van Aelst 2017).

A germane topic within scholarship analyzing the relationship between media and politics has been the media mix that electoral candidates and their teams use to convey their messages in a context that Andrew Chadwick (2017) has characterized as a hybrid media system. This has evolved throughout modern history (Seidman 2008). In the twentieth century, a key turning point in this regard within the Global North had to do with the emergence of the televised presidential debate as a key institution for showcasing candidates to their voters (Druckman 2003). The 1960 debate between Richard Nixon and John F. Kennedy at WBBM studios in Chicago, the first televised debate in the United States, broke new ground in the repertoire of campaign strategies. *New York Times* reporter Jack Gould described it as "a dignified and constructive innovation in television campaigning. Undoubtedly it helped to quicken public interest in the Presidential contest."[11] Nixon and his team famously dismissed the importance of caring for the candidate's image on the television floor—which some argued might have contributed to losing the debate to a young Kennedy, more skilled at performing for the small screen.[12]

The electoral campaign run by former president Barack Obama in the United States in 2008 marked another turning point (Johnson and Perlmutter 2011; Bimber 2014; Chadwick 2017): It made it clear that the world of politics could make use of platforms such as Twitter—and fourteen other platforms, according to Wikipedia[13]—to mobilize parts of the electorate.

To examine how candidates imagine the relationship between traditional and social media in presidential campaigns, Luc Chia-Shin Lin (2016) looks at the case of the presidential elections of Taiwan in 2012. He interviews people in either campaign staff, journalism, or political communication research positions. The question guiding his work is whether the growing popularity of the Facebook pages of election candidates alters the relationship between mass media and platforms during election campaigns. Lin finds that Facebook was of particular significance to both candidates and journalists. Candidates themselves "attempted to view their Facebook pages as headwaters of mass media; this view allows social media to operate as an intermediary between candidates and mass media" (Lin 2016, 211). Journalists, likewise, perceived Facebook as a source of news about the candidates. For instance, Facebook posts from candidates during the 2012 Taiwanese presidential campaign would in some cases be published before press releases were sent to traditional media. Ultimately, Lin (2016) observes a "parallel" relationship between social and traditional media during election campaigns: "From the perspective of journalists, this parallel relationship exemplifies the frame contest and enables them to perceive candidates' strategic purposes. Yet, from the perspective of the candidates, the parallel relationship points to a need to increase the influence of their Facebook pages because the pages' popularity indicates how online and off-line environments intertwine with each other" (208).

Presidential candidates use a mix of traditional and social media to introduce themselves to their electorate and convince them to change their vote because it is assumed that the media have a certain degree of power to modify the behavior of their publics (Gerber, Karlan, and Bergan 2009). But far from having a direct impact, scholarship has shown that they influence the formation of opinion in ways that are not necessarily linear or self-evident (McCombs, Shaw, and Weaver 2014; Benkler, Faris, and Roberts 2018). As the canonical article by Maxwell McCombs and Donald Shaw (1972) puts it, traditional media are not necessarily good at telling us what to think but about which issues to think; in other words, print and broadcast media have a certain degree of power in setting the agenda (Cohen 1963; Brosius and Kepplinger 1990; Vliegenthart and Walgrave 2008; McCombs and Valenzuela 2021). But what is the capacity of social media platforms to shape public opinion and set the agenda of their users? In addition, is it possible to understand this capacity in isolation, without relating it to that of traditional media?

Samuel Mochona Gabore and Deng Xiujun (2018) study how journalistic frames from online news impact public opinion expressed on social media. They undertake a comparative content analysis focused on the coverage of the construction of the Addis Ababa–Djibouti railway line in Ethiopia—which was considered to be, at the time of the study, the first modern railway in East Africa. Specifically, the authors compare traditional media coverage and Facebook posts. In doing so, they find a pattern whereby traditional media influences the frames and issues discussed on Facebook: "[S]ocial media users are affirming or criticising issues in a similar tone as they are presented by traditional media. This implies that evaluative opinions of social media users are formed as the result of exposure to news media's labelling of issues" (Gabore and Xiujun 2018, 35).

However, Gabore and Xiujun note a discrepancy within that trend that is worth mentioning: traditional media coverage with a neutral tone show a nonsignificant relationship with posts published on Facebook. This, to the authors, indicates that "information presented in neutral tone has weak influence on social media opinion formation" (Gabore and Xiujun 2018, 35). As we suggested in chapter 2, there has been considerable interest in investigating the relationship between political content and ideological polarization (Stieglitz and Dang-Xuan 2013; Benkler, Faris, and Roberts 2018; Fletcher, Cornia, and Nielsen 2020). Although different studies have yielded dissimilar results (Bondes and Schucher 2014; Johnson 2018), the approach of Gabore and Xiujun (2018) is of particular interest because it illustrates how one aspect of the phenomenon—which factors contribute to potential polarization in social media—acquires greater clarity when examined from a cross-media perspective.

In the two studies we discussed in this section, a comparison between traditional and social media enabled the analysis to make more visible and understandable communication dynamics that would otherwise have remained less visible and intelligible. The work of Lin (2016) comparing the media mix in an electoral campaign showed that a parallel relationship between traditional and social media emerged, whereas Gabore and Xiujun (2018) indicated that traditional media had relatively more capacity to frame coverage than a social media platform.

Approaches

Two common alternative ways of approaching the comparison between traditional and social media have been emphasizing either continuities or discontinuities. The first approach is partly based on the idea that traditional and social media are not only part of a historical continuum, but they can also be thought of as complex sociotechnical artifacts that belong to a single information ecology of mutual influence (Dubrofsky 2011; Hermida 2014; Chadwick 2017; Humphreys 2018). The second approach is premised on the notion that traditional and social media can be examined separately by contrasting their capacity to affect one or more outcome variables (Sayre et al. 2010; Stefanone, Lackaff, and Rosen 2010; Kalsnes, Krumsvik, and Storsul 2014; Valenzuela, Puente, and Flores 2017).

One example of approaches emphasizing continuity can be found in a study by Robin Rymarczuk (2016), who links discourses of resistance to social media to the resistance to the landline telephone in the early twentieth century. The literature on non-use has a long history in communication studies and science and technology studies (Fischer 1992; Kline 2000; Wyatt 2003; Foot 2014; Syvertsen 2017; Hesselberth 2018). It reveals not only alternative forms of reappropriation of media artifacts but also ways in which identities are constituted around the rejection of technology. Rymarczuk explains that "the intrusion that social media makes on the individual, be it user or non-user, has added a layer of complexity to daily life comparable to the decision to whether to answer the phone or not in the 1900s" (2016, 46).

Through an archival analysis of American, British, and Dutch press between the late nineteenth and early twentieth centuries, the author examines the alleged intrusions of privacy generated by the emergence and subsequent massification of the telephone. He argues that "The reason that the collapse of social relations because of technology is thought of as a new issue, is because different people, experts and fields of science speak loudest on the subject today. These arguments and concerns are, however, not new at all. They are just repackaged by contemporary paradigms. The state of resistance to social media is certainly evidence for the fact that the non-user of old wasn't heard accordingly: because contemporary concerns reign, hardly impacted by those same early arguments for non-use" (Rymarczuk 2016, 47)

An alternative approach to the comparisons between traditional and social media has been to highlight areas of discontinuity. In the context of the Arab Spring (Lotan et al. 2011; Wolfsfeld, Segev, and Sheafer 2013; Kraidy 2016), what types of narratives and discourses did circulate in print journalism versus on Twitter? Stefanie Ullmann

(2017) aims to answer this question through a discourse analysis of articles from six newspapers in the United States, the United Kingdom, and the MENA (Middle East and North Africa) region, and a sample of 1,000 tweets which, during the days of protest in January 2011, had used the hashtag #Jan25. Distinguishing between the ways in which demonstrators, the police, security forces, and the Egyptian government were talked about, Ullmann finds that "While there does exist a certain lack of clarity or even ambiguity in the portrayal of police forces and their behaviour in mass media, . . . the tweets display a clear tendency to portray the police as the weaker entity that is unwillingly overwhelmed by and thus inferior to the protesters" (2017, 175).

The author observes, then, significant differences in the discourses that circulated within traditional and social media in the face of a phenomenon of the magnitude of the Arab Spring, supporting the notion that "when compared to mass media, social media may contain ideologies that are less institutionalized, while at the same time enabling the limitless expression of political and social opinions" (Ullmann 2017, 166).

In this section we reviewed two alternative approaches to cross-media scholarship. In both cases a comparative sensibility elicited findings that would have been less visible in platform-only accounts: the historical continuities in representations of the rejection of new technologies, as Rymarczuk (2011) showed, and the lower levels of institutionalization in protest discourse on Twitter versus traditional media, as illustrated by Ullmann (2017).

Methods

There is a productive distinction that organizes methodological matters in comparative cross-media work and that also relates to the continuity–discontinuity pair mentioned in the previous section: the distinction between diachronic and synchronic approaches to communication phenomena. While diachronic or longitudinal methodologies underscore the importance of attending to the passage of time, the synchronic or cross-sectional counterparts focus on a given phenomenon at a particular moment. Both methodological strategies have been deployed in cross-media comparative scholarship: qualitatively through in-depth interviews, discourse analysis, and focus groups, among others (Lin 2016; Törnberg and Törnberg 2016; Schmidt et al. 2019); and quantitatively through surveys, experiments, and social network analysis, among others (Kwak et al. 2010; Schultz, Utz, and Göritz 2011; Abdenour 2017).

A generative implementation of a diachronic research design is undertaken by Sebastián Valenzuela, Soledad Puente, and Pablo M. Flores (2017) to examine the relationship between the agenda of traditional and social media in the coverage of the earthquake that occurred in Chile on February 27, 2010, and that resulted in more than 500 fatalities. Just as Lin (2016) conceived of news and platforms as parallel media systems, and Gabore and Xiujun (2018) were interested in understanding the agenda-setting power of online news over social media, Valenzuela and colleagues compare the evolution of topics covered by journalists on broadcast television and Twitter during the first week after the earthquake. In the case of television, they analyze the content of the newscasts in the country's most important networks during the prime-time slot; for Twitter, they look at a sample of 270 messages produced by journalists working in Chilean media. Their research design seeks to counter the trend whereby "most published research takes a platform-centric perspective, in which the impact of Twitter on journalistic practice and news coverage is studied in isolation from other media" (Valenzuela, Puentes, and Flores 2017, 616).

Contrary to the findings of Gabore and Xiujun (2018), their attention to dynamics happening over time enables them to show "a reciprocal but asymmetrical relationship in which television news shows are more likely to 'adopt' the issue agenda of journalists' on Twitter than vice versa" (Valenzuela, Puentes, and Flores 2017, 631). It is thanks to their perspective, focused on how the media agenda evolved over the course of one week, that the authors are able to identify inter-media dynamics between traditional and social media. According to Valenzuela and colleagues, "this is consistent with prior evidence that online platforms, including social networks, discussion forums, and search engines can influence news coverage of traditional media" (2017, 631).

A fruitful example of a study adopting a synchronic research design instead is that of Rebecca Nee and Valerie Barker (2020), about the social impact of coviewing in situations of second screening. The authors define the phenomenon of second screening as "using another device (laptop, cellphone, tablet) to text, go online, or use social media in a complementary manner to what is being watched on television. . . . Second screening implies that the viewer's focus is on both screens simultaneously" (Nee and Barker 2020, 3). This practice, which since 2013 has been measured by the Nielsen rating system, has been associated with younger age groups and in many cases is referred to as "social television" (Chorianopoulos and Lekakos 2008; Giglietto and Selva 2014; Selva 2016; Wohn and Na 2011). Nee and Barker draw upon surveys with teenagers and university students in Qatar and the United States in 2017 and 2018 to examine cases of second screening with both traditional television and streaming services. They are interested in understanding, among other things, whether the experience of consuming content in this way ends up being "lonely" by force or whether social benefits can arise from consuming content via YouTube and television while also using other platforms or messaging services to communicate with people who are also consuming that same content physically apart from each other. Nee and Barker (2020) find that second screening "promotes a sense of community for users in both contexts, even if the viewing is not taking place simultaneously with others. Although some differences were found based on age, gender, and ethnicity in second screening, the most surprising results are not the differences, but the similarities of co-viewing outcomes for both traditional television and streaming services" (13).

Regarding potential social benefits the authors note that for both traditional television and streaming services, such as YouTube, "even when people are using another device without the intention of communicating with others, they could be achieving a sense of community as a byproduct of their search for information about the show" (2020, 14).

In this section we considered two typical methodological strategies regarding the role of temporal matters that converge in showing the descriptive and explanatory potential of comparative work. Because Valenzuela and colleagues (2017) compared the agenda-setting power of both television broadcasting shows and posts on Twitter over time, they were able to illuminate inter-media dynamics that would have been otherwise left opaque. Since Nee and Barker (2020) contrasted the social effects of coviewing in second screening practices between television and YouTube, they shed light on the fact that social media could also produce prosocial effects—contrary to the idea of smartphones being isolating.

Interpretations

Scholars have enacted several interpretive frames to make sense of findings obtained from cross-media comparisons. We underscore two common ones: reinforcement and displacement.

According to the idea of reinforcement, not only do social media present a logic with antecedents in traditional media, but also both types of media mutually shape each other in ways that end up creating feedback loops. As we suggested in the opening vignette about the Kardashian-Jenner clan, there seems to be a relationship between the culture of selfhood conveyed in early reality television shows, on the one hand, and the cult of self-image and the daily accounting of the self that is part and parcel of platforms on the other hand (Stefanone and Lackaff 2009; Kraidy 2009; Marwick 2013; Khamis, Ang, and Welling 2017; Psarras 2020). These platforms, in turn, end up shaping the ways in which content is produced and formats are designed for traditional media. This can be seen not only for entertainment but also for news because journalists increasingly source and communicate on social media (Hedman and Djerf-Pierre 2013; Canter 2015; Brems et al. 2017; Mellado and Hermida 2021). Sometimes these transformations occur at the intersection of entertainment and news, for instance, when articles about the passing of a celebrity are filled with the reactions on social media by other celebrities, a dynamic which we will elaborate further in chapter 5.

The reinforcement relationship between the logic of reality television formats and social media such as Facebook is the subject of a study by Rachel E. Dubrofsky (2011). According to the author, reality television programs seem to have fostered subjectivities that support surveillance schemes, in which the life of the subject is placed at the service of consumption and presentation in front of mass audiences. Thus, in reality television "participants are habituated to putting the self on public display for entertainment purposes" (Dubrofsky 2011, 124). Furthermore, in the case of Facebook, Dubrofsky argues, it has become routine to resort to "using surveillance technologies in the service of producing consumable products (bits of data), suggesting the desirability of living a life that can withstand being under surveillance, as well as a life that can be broadcast to an audience" (Dubrofsky 2011, 124).

Finally, Dubrofsky (2011) finds important differences between reality television and Facebook: "[T]he hands-on shaping of the reality television subject by television workers differs from Facebook's processes of subjectification. On Facebook, users largely mediate their own subjectivities without third-party intervention" (Dubrofsky 2011, 117).[14] The discussion of the relationship between information and communication technologies and surveillance, however, is far from settled. A prominent example of the ongoing relevance of this open-ended debate is Shoshana Zuboff's *The Age of Surveillance Capitalism* (2019). In this book, Zuboff proposes that the production logic of the world's most successful technology companies and platforms is based on a scheme of surveillance and expropriation of personal data. Complementing this examination with comparative perspectives could certainly shed light on the institutional histories of modern media that provided key conditions for this logic to emerge and consolidate.

The interpretive frame of displacement when it comes to cross-media studies can be seen, for instance, in research about the growing reliance on social media for seeking information about current affairs (Zhou et al. 2019; Lewis 2020). Sayre et al. (2010) study the coverage of Proposition 8 in California during 2008 and 2009. Proposition 8 was a referendum to constitutionally repeal the right to same-sex marriage; it was passed in the California state elections of November 2008 and was afterward overturned in court.[15] Sayre and colleagues compare the press coverage of California media, news indexed by Google News, and YouTube videos. Their aim is to understand the potential mutual influence between traditional and what they call "online media" in terms of opinion formation and agenda setting. Their ultimate interest is "the question of whether these new social media forums produce different agenda-setting cues than those the public is already exposed to in other, more established media" (Sayre et al. 2010, 15). To this end, they track content mentioning Proposition 8 across newspapers, Google News, and YouTube videos during a period of fourteen months between 2008 and 2009. Sayre and colleagues examine these data via a time-series analysis, concluding that "It was opponents of Proposition 8 who accounted for nearly all of the activity on YouTube following the election. . . . YouTube was being used as a platform for people to register opinions that they felt were not being represented in the mainstream" (2010, 24).

They add that this "is symptomatic of a traditional media system that may be losing some of its agenda-setting ability to emerging social media" (Sayre et al. 2010, 26).

This section focused on two common interpretive frames to make sense of comparative cross-media dynamics: reinforcement and displacement. Beneath the differences, an issue that cuts across the studies surveyed is the power of comparative analysis to question assumptions such as that social media are the first technological artifacts to impose constant exposure of the self, in the case of Dubrofsky (2011), and to uncover the capacity of YouTube to work as a political space for self-expression that is unparalleled by mainstream media, as shown by Sayre and colleagues (2010).

Conclusions

We opened this chapter with a vignette about a reality television program in which the strategic management of scoops reveals a web of connections across traditional and social media. We paired it with a second vignette about the reconfiguration of the genre of political commentary in Twitch to engage with audiences in ways that appeal to those not interpellated by how journalists present this content in newspapers and television. Then, we discussed eight studies that illustrate the descriptive fit and explanatory power of scholarship that aim to make sense of social media practices in relation to those that are typical of traditional media. The selected studies dealt with a range of historical periods from the nineteenth century to the twenty-first century, spanning several key moments of the twentieth century. They also examined different media and communication technologies, from the landline telephone to Twitch, including television, Facebook, and YouTube.

In their diversity, the vignettes and studies highlighted the relevance of cross-media comparative work to better understand social media phenomena and to question approaches that tend to overde-termine the power of one technology just because it is "newer" than others (Czitrom 1982; Marvin 1988; Bolter and Grusin 1999; Gitel-man 2006; Peters 2009; Bourdon 2018). The comparative perspective thus helped to discern significant similarities and differences that otherwise would have been less visible—or perhaps altogether invis-ible. As with the studies discussed in chapter 2, the breadth of the methodological and theoretical strategies pursued by the authors indicates that the comparative turn advocated in this book consists of an epistemological stance flexible enough to accommodate a wide range of approaches regarding how to gather and process data as well as explain the potential variance in the findings.

In a historical media context characterized by what Henry Jen-kins (2006) calls "participatory culture" and "transmedia storytell-ing," a comparative cross-media approach allows us to move toward more relational and holistic views that ultimately enable us to refine our understanding of social media. Our perspective is con-sistent with the arguments that media theorists have repeatedly made since digital and networked information technologies began their ascent in everyday life in the mid-1990s to better analyze the relationship between older and newer media (Manovich 2002; Jen-kins 2006; Chun 2008; Gitelman 2006; Hayles 2007; Peters 2009). As we noted in chapter 1, Jay David Bolter and Richard Grusin's (1999) relational theorizing about the dynamics of remediation underscores a central point emerging from this chapter: the con-tinued relevance of traditional media in establishing the geneal-ogy of what we associate with the new—in this case, social media platforms—and the persistent influence of what we conceive of as old. For instance, we observed the influence of television aesthetics, in particular reality television, in common modes of information presentation on platforms, where the cult of the self coexists with strategic image management and standardized forms of surveil-lance, in a logic of constant production of aestheticized selfhood

(Marwick 2013; Duffy 2017; Brydges and Sjöholm 2019; Arriagada and Ibáñez 2020).

Our approach builds upon theorizing of both cultural and institutionalist lineage. While Henry Jenkins (2006), as noted in chapter 1, characterizes convergence culture in terms of content flow, industry cooperation and audience behavior, Andrew Chadwick's "hybrid ontology" to understand the contemporary media system "eschews dichotomous modes of inquiry and instead invites us to focus on the overlaps and the in-between spaces that open up between older and newer media technologies, genres, norms, behaviors, and organizational forms" (Chadwick 2017, xii). What is especially evocative in the context of this chapter is how both theorists, while arguing from divergent traditions of inquiry, nonetheless converge on the idea that analyses of social media isolated from their relationships with traditional media and the wider social environment lack both descriptive fit and heuristic power. That is, they fail to illuminate the everyday practices of users and to account for variations across them. This is because, as we saw throughout this chapter, people appropriate platforms often in relation to traditional media practices—even the absence of the latter provides relevant information, as in the case of Piker's coverage of presidential debates on Twitch. It is through comparative perspectives that this media coexistence and the particularities that mark the uptake of each medium can begin to be foregrounded in the analysis.

Building upon cross-national and regional, and cross-media comparisons, research also shows that users appropriate a given social media platform in relation to other platforms (Zhao, Lampe, and Ellison 2016; Boczkowski, Matassi, and Mitchelstein 2018; DeVito, Walker, and Birnholtz 2018; Valenzuela, Correa, and Gil de Zúñiga 2018; Tandoc, Lou, and Min 2019). Although we have so far used the term *social media* to refer generically to the set of platforms available, in the next chapter we turn our gaze to scholarship that has compared across platforms, thus further underscoring the inherent plurality of social media technologies.

4

Cross-Platform Comparisons

Introduction

Platforms in the Age of Technical Reproduction

In mid-November 2020, the official Twitter account publishes a tweet announcing the creation of Twitter Fleets, a feature designed to communicate "that thing you didn't Tweet but wanted to but didn't but got so close but then were like nah."[1] The publication is quickly followed by thousands of tweets from around the world that take the announcement as a source of humor, mostly in the form of memes.[2] The humor, which blends mockery, frustration, and amusement, foregrounds the clear similarity between Fleets and a feature already present in platforms such as Snapchat, Instagram and Facebook: stories. Both Fleets and stories offer a brief audiovisual format, also ephemeral by default, which invites users to share the here and now of their lives.

It turns out that in the current instantiation of our age of digital reproduction (Shifman 2007; Knobel and Lankshear 2008), imitation—paraphrasing Oscar Wilde—might be the sincerest but not the most appreciated "form of flattery that mediocrity can pay to greatness." Users do not hide their disappointment at what appears to be an almost unbearable loss of Twitter's aura. *Aura*, in Walter Benjamin's (1936) terms, refers to the uniqueness and permanence of original works of art, two features challenged by mass reproduction and copying.[3] Twitter users, seemingly upset, describe in a cynical key the fatigue of finding the same functionality replicated time and again across the social media landscape. Moreover, the audiovisual aesthetics of Fleets seem to transgress the spontaneous and, above all, written culture that audiences usually attribute to Twitter (Burgess and Baym 2020).

One of the complaints most circulated on social media in the aftermath of Twitter's presumed imitation paradoxically links to another imitation. It takes up the internet meme "will now have stories,"[4] which emerged a few years ago as a reaction to Facebook's 2017 decision to include stories on its platform. The capability had already been added to WhatsApp that year, and to Instagram during 2016, as a way of competing with the popularity of the feature deployed by Snapchat in 2013.[5] The "will now have stories" meme superimposes images emulating stories on various everyday objects—from a banana to a pregnancy test, and from a calculator to a McDonald's menu—in order to mock the seeming lack of originality of the social media realm. In November 2020 the meme is resurrected to allude to the new wave of cross-platform copying, this time led by Twitter. One of the jokes, posted by a Twitter user, includes the "pointing gun meme," where the characters of the television series *The Office* point to one another, and states, "Tik Tok copied Vine, one of Twitter's biggest failures. Instagram copied Tik Tok, by making Reels. Twitter copied Instagram, by introducing Stories, which Instagram stole from Snapchat."[6] The user adds a Twitter thread, "ME, realizing that if all of these apps cannibalize themselves and make terrible product updates that make us want to use them less, we might all get our freedom back."[7] Less than a year after the announcement of its launch, Twitter closed down Fleets with the following announcement: "we're removing Fleets on August 3, working on some new stuff we're sorry or you're welcome."[8]

Lowest Common Denominator

On March 15, 2019, a terrorist attack is perpetrated in two mosques located in Christchurch, New Zealand. A white supremacist and conspiracist murders fifty-one people and injures forty more innocents. Before doing so, he sends an online manifesto via email to thirty recipients and posts links to it on Twitter and 8chan, with the aim of making the impending massacre go viral. The horror is magnified when the killer decides to broadcast the first shooting on Facebook Live for seventeen minutes; the video then remaining on his profile.[9] On platforms whose core offer has to do with content moderation, as argued by Tarleton Gillespie (2018), the transmission of the attack momentarily dodges the human and algorithmic controls that Facebook enforces around the globe. A *New Yorker* article explains that before Facebook's specialized team removed the content, the video had already been viewed by 4,000 people and not even one had reported it until 29 minutes after the start of the live transmission.[10]

In the wake of the events, various officers of the New Zealand government, including Prime Minister Jacinda Ardern, and other world leaders, such as French President Emmanuel Macron, urge companies—especially American ones that run platforms such as Facebook, Twitter, Reddit, and YouTube—to show accountability for their policies of moderating violent and hateful content.[11] Two months later, a meeting of heads of state overseen by Ardern and Macron is held in Paris "to respond swiftly and effectively in the event of a terrorist attack and/or of viral terrorist content online."[12] The objective is both normative and ethical. It is posited that a common, international standard of transparent policies to counter hate crimes could help prevent them in the future. It also explains that clear moderation policies, as well as honoring the right to be forgotten, would protect the memory of victims whose images were still circulating on platforms days after the event. A report entitled "Antisocial media," produced by New Zealand think tank The Helen Clark Foundation, recommends that "The Government meet with social media companies operating in New Zealand to agree on an interim Code of Conduct, which outlines key commitments from social media companies on what actions they will take now to ensure the spread of terrorist and other harmful content is caught quickly and its further dissemination is cut short in the future. Limiting access to the livestream feature is one consideration, if harmful content can genuinely not be detected."[13]

In the news about the aftermath of the attack, different traditional media organizations, from *Le Monde* to *Vice*, begin to cover how each platform proposes to solve ad hoc moderation issues.[14] This ends up revealing that it is probably not correct to assume that all platforms share a common denominator when it comes to moderating violent and hateful content. Certain idiosyncrasies in the moderation of platforms that are usually kept out of the limelight become suddenly exposed. The similarities and differences between them turn into a truly significant issue for the public and the polity at large.

How Comparisons across Platforms Matter

The contrast between the two preceding cases is striking. On the one hand, we describe developments around a relatively minor technical capability and the satirical reactions it triggered. On the other hand, we revisit a tragic event of major gravity and the stern international reaction that followed. In spite of their evident differences, both cases share an issue that constitutes the central node of this chapter: the descriptive, explanatory, and interpretive gains that arise from comparisons of practices and discourses across platforms. This epistemic operation exposes, in the case of the failed Twitter Fleets and its successful Snapchat, Instagram, WhatsApp, and Facebook predecessors,

a growing technical homogeneity across seemingly divergent platforms. As we saw, this homogeneity is quickly picked up by users who
complain and express themselves sometimes humorously against
what they see as a lack of originality across platforms. They thus
argue for a clear distinction in the imagined and inhabited cultures
that populate the social media landscape. In contrast, the events following the Christchurch mosque shootings make visible the heterogeneity that also exists across platforms in other dimensions. When
numerous government authorities, on behalf of their respective citi-
zenries, call for common standards and international mechanisms to
control hateful and violent content, a lack of homogeneity and stan-
dardization of regulatory practices across social media is exposed.

As different as they are, both stories point to a shared matter, that
is, they can be best understood through a comparative lens. Users
view a given platform's decision to adopt a feature as either innova-
tive or not, depending on their knowledge about decisions made by
other platforms. Governments and citizens demand explanations
from platforms about their preventive measures and moderation
mechanisms regarding hateful and violent content because they
assume that they are likely to have different responses or, at least,
that their infrastructures do not necessarily respect a common—let
alone international—standard.

Comparing across platforms is grounded in our everyday experiences. As noted in chapter 1, worldwide the average social media
user has an account on more than seven platforms (often using more
than one concurrently and relationally) with different sociodemo-
graphic groups adopting different platforms and/or combinations
of them (Hargittai 2007; Hargittai and Hsieh 2010; Horvát and Hargittai 2021; Matassi, Mitchelstein, and Boczkowski 2022). People
often sense that certain ways of communication and self-presentation
are socially acceptable on some platforms and not on others (van
Dijck 2013; DeVito, Birnholtz, and Hancock 2017; Scolere, Pruchniewska, and Duffy 2018; Duffy and Chan 2019). They also perceive

that certain posting frequencies or criteria for reacting to content are more appropriate on some platforms but not all (Kaun and Stiernstedt 2014; Bayer et al. 2016; French and Bazarova 2017; Boczkowski, Matassi, and Mitchelstein 2018). Twitter users' reaction to the incorporation of Fleets indicated that something did not feel right: The functionality broke a certain implicit, but nonetheless powerful, norm about the types of content suitable for it. Making sense of these situations invites comparisons that interrogate both obvious cross-platform practices as well as practices that do not take place in one or more of them because of usually unstated yet highly consequential social conventions.

In the following pages we will present eight studies that deal with key issues in cross-platform comparative research. As with the previous two chapters, we will organize them according to topics, approaches, methods, and interpretations. Finally, we will conclude with an analysis of how comparative work allows us to unpack the concept of social media into its main constituents instead of treating it as a homogeneous whole. Inspired by traditions of inquiry that propose relational and holistic views, we will argue that when we speak of platforms it is important that we imagine and study them in their interconnected plurality. If, as Lisa Gitelman has argued, traditional "media *are*" (2006, 2; emphasis added), then it is reasonable to expect that social media also should be understood as plural entities.

Topics

Two topics recurrently addressed by cross-platform scholarship are the presentation of the self and the impact of social media on mental health.

Research on the ways in which people present their selves in private and public environments has a long tradition in microsociology (Blumer 1969; Knorr-Cetina 2009; Benzecry and Winchester 2017), which has then permeated into communication studies, especially through the influence of Erving Goffman's work (1959, 1967). According to these traditions of inquiry, social interaction is a space in which intersubjective meaning is produced and social order is built. The presentation of the self is an important aspect of that process (Marshall 2010; Marwick and boyd 2011; Litt 2012; van Dijck 2013).

The selfie, which was named word of the year by *Oxford Dictionaries* in 2013,[15] has been one of the most prevalent genres of self-presentation in everyday uses of social media (Katz and Crocker 2015; Marwick 2015; Chua and Chang 2016). Interested in the "conversational capacity" of selfies, Stefanie Duguay (2016) compares model and actress Ruby Rose's self-presentation on Instagram and Vine. Using the walk-through method of analysis of the discourses surrounding these apps, the author seeks to understand the production and circulation of different forms of LGBTQ visibility in the selfies that Rose shares on both platforms. To Duguay, "the conversational capacity of LGBTQ people's selfies, as performances of sexual and gender identities . . . influences the potential for circulating counter-discourses and forming queer publics" (Duguay 2016, 3).

According to Duguay, while certain selfies tend to reproduce heteronormative gender stereotypes, others counter hegemonic discourses on gender and sexuality. Duguay focuses on three parameters: "range, the variety of discourses addressed within a selfie; reach, the circulation of selfies within and across publics; and salience, the strength and clarity of discourses communicated through a selfie" (2016, 2). She chooses to compare Instagram and Vine because they share an emphasis on visuality as well as other characteristics, including having been launched or bought by popular platform companies and presenting a similar set of technical functionalities. From a detailed analysis of the discourses surrounding the description of the applications in mobile applications stores, and of images shared by Rose, Duguay concludes that "Although Instagram provides many content generation tools, its aesthetic formula decreases the salience of counterdiscourses in selfies, while Vine's scarcity of tools leaves room for users to increase the salience themselves. Without a layer of editing or filters, Viners' personal aspects become salient, making identity discourses prominent and available for conversations across publics" (2016, 9).

A second theme that appears frequently in cross-platform studies is the impact of social media on mental health. There has been significant public concern about the effects (generally seen as negative) that the adoption of platforms can produce (Twenge et al. 2018; Orben 2020; Vanden Abeele 2020). Within this context scholars inquire into whether platforms affect preexisting states, such as loneliness (Hunt et al. 2018); encourage *de novo* mental health conditions, such as eating disorders (Saunders and Eaton 2018); and/or whether certain individual characteristics, such as depression, lead to the use of the platforms in the first place (Ozimek and Bierhoff 2020).

Sonja Utz, Nicole Muscanell, and Cameran Khalid (2015) examine the experience of feelings of jealousy in the context of romantic relationships and their ties to the use of Facebook and Snapchat. The comparison between these two platforms is partly informed by the notion that Snapchat is used more for intimate communication among youth than other platforms due to the ephemerality of its content (Boczkowski 2021). Facebook, on the contrary, is usually associated with a more public communication culture, where posts often stay on the news feed and the boundaries among the different audiences of each user collapse more easily (Marwick and boyd 2011; Bayer et al. 2016; Litt and Hargittai 2016). Drawing from an online survey with participants in various European countries, the authors inquire into the motivations for using each one of these platforms as well as feelings of jealousy associated with their use.

Regarding the issue of motivations, Utz, Muscanell, and Khalid (2015) observe that even though "Snapchat use resembles Facebook use in many respects, . . . Snapchat was used somewhat more for flirting than Facebook" (2015, 144). Concerning the level of jealousy experienced by users of both platforms, "although both media did not trigger extremely high levels of jealousy, Snapchat did elicit more jealousy than Facebook" (Utz, Muscanell, and Khalid 2015, 144). They explain, however, that "when it comes to receiving (vs. sending) a post from an unknown potential rival, jealousy was higher on Facebook. It seems that threats from third persons are perceived as more threatening when they are public" (145). To the authors, this reveals that "although social media can evoke jealousy, they do not make everyone highly jealous" (145).

In both topical examples examined in this section, cross-platform comparisons were key to illuminating the dynamics under study either by showing how the circulation of discourses around LGBTQ experiences can significantly differ according to the platform at stake (Duguay 2016), or by shedding light on the idea that not all platforms are similarly associated with certain psychological states (Utz, Muscanell, and Khalid 2015). None of the dynamics that apply to individual platforms would have been made visible without accounts that interrogated relationships with other platforms.

Approaches

Two types of approaches dominate cross-platform scholarship—what we call linearity versus circularity. The first, and most common, contrasts the impact of either an independent variable on two or more platforms, or two or more platforms on a dependent variable. The second, although less frequent than the first, examines relationships across platforms.

A study by Sebastián Valenzuela, Teresa Correa, and Homero Gil de Zúñiga (2018) provides an illustration of the first approach. This work takes up discussions that have appeared in previous chapters concerning the relationship between political participation and social media use (Bennett 2012; Gil de Zúñiga, Jung, and Valenzuela

2012; Boulianne 2015). Valenzuela, Correa, and Gil de Zúñiga's (2018) goal is to understand the connections that cut across political participation, political information consumption, and social media use for the cases of Twitter and Facebook. The more specific question that guides their inquiry is whether any of the affordances of these platforms—and the social relationships they activate, categorized in terms of the distinction between weak and strong ties—are conducive to specific forms of political participation. They examine this matter through the analysis of a survey of young Chileans conducted in 2014.

Valenzuela and colleagues identify important differences between both platforms. Whereas Facebook allows for a rather symmetrical and reciprocal connection between users, Twitter offers the possibility of relations that might be asymmetrical or unidirectional—with one party not having to necessarily accept the "follow" request from the other one. Thus, "both social media platforms have positive effects on mobilizing Chilean citizenry, and fostering political protest behaviors. However, these relationships emerge from distinct social network structures within these social media platforms. On the one hand, results indicate that on Facebook, strong-tie connections are conductive to further protest behavior, while exposure to weak ties conveys a much weaker influence on this type of political activity. Conversely, weak-tie connections in Twitter seem to lead people to engage in protest behavior; interactions with strong ties on this medium have no discernible impact" (Valenzuela, Correa, and Gil de Zúñiga 2018, 128–129).

The second type of approach, that of circularity, has concentrated on cross-platform relations. The Cambridge Analytica scandal of 2018 (Susser, Roessler, and Nissenbaum 2019; van Dijck 2020) revealed the existence of an ultra-targeted strategic communication apparatus based on an ecology of misinformation traveling from one country to another (Walker, Mercea, and Bastos 2019). Since then, many public and news media discussions have emerged around issues of fake news, misinformation, and disinformation. These topics are not new (Jaramillo 2006; Boczkowski and Mitchelstein 2021; Seo and Faris 2021), but in recent years there has been an explosion of scholarship triggered by events such as the Brexit vote in the United Kingdom and the 2016 presidential election in the United States, to such an extent that the term "fake news" was named word of the year by *Collins Dictionary* for 2017.[16]

Josephine Lukito's (2020) work seeks to shed light on the digital infrastructure behind the systematic plan to disseminate false information during the 2016 US presidential election. The author examines the disinformation campaign strategized and executed by the *Internet Research Agency* from 2015 to 2017; also known as IRA, this has been linked to actors with ties to the Russian government (Polyakova 2019). Lukito focuses on understanding the coordinated manner in which the campaign was deployed on Facebook, Reddit, and Twitter. She argues that the multiplatform logic of the campaign responds to the fact that a greater number of platforms operating in unison can potentially increase the frequency with which a user is exposed to fake news. She explains that "while previous studies have looked at the dynamics of this campaign on individual social media platforms (e.g., Broniatowski et al. 2018), none have empirically tested the possibility that multi-platform disinformation campaigns are internally coordinated" (Lukito 2020, 239).

In her analysis, Lukito (2020) distinguishes between paid content, which has to do with positioning a post through the purchase of advertisement, and organic content, which arises via word-of-mouth interactions and/or unpaid recommendations. She suggests that paid content via Facebook ads will happen on a different timeline than organic content on Twitter and Reddit. More importantly for the purpose of this chapter, Lukito hypothesizes that the dissemination of content on one platform might inform and influence the dissemination of content on the others. She undertakes a time-series analysis of the data that Facebook, Reddit, and Twitter released to the public after the Cambridge Analytica scandal broke and its subsequent treatment in the US Congress.

Lukito (2020) finds that paid Facebook ads had no temporal relationship with Reddit and Twitter content, and that the relationship between Reddit and Twitter was unidirectional in the sense that posts on the former influenced those on the latter. Her explanation is that Reddit might have been used to test the effectiveness of a piece of content before reinforcing it on Twitter. This platform, Lukito argues, may have been more relevant than Reddit because of its privileged place within journalistic practice (Hermida 2010; Lasorsa, Lewis, and Holton 2012; Barnard 2016). Lukito concludes that "strategic communicators—including the Internet Research Agency—use many platforms in tandem to spread and reinforce messages. It therefore behooves scholars to study political communication in a multi-platform context, rather than looking only at messages within one platform" (2020, 250–251).

In the two studies presented in this section, the comparative stance proved central for shedding light on the phenomena at hand. Whereas Valenzuela, Correa, and Gil de Zúñiga (2018) found through comparison of Facebook and Twitter that the relationship between social media use and political participation was significantly moderated by the social networks that the user activates, Lukito (2020) was able to show the circulation of content from one platform to another within a process of orchestrated disinformation. Had these two papers focused on one single platform isolated from the others, they might not have been able to properly identify significant variations or mechanisms in either political participation or flows of disinformation, respectively.

Methods

Cross-platform research has often used quantitative techniques, such as online surveys and computational methods. To a lesser extent, some studies have utilized mixed quantitative and qualitative methods, such as surveys with focus groups or interviews; others have engaged purely qualitative tools.

In 2015, at an Australian Football Association (AFL) game, Adam Goodes, a player of Adnyamathanha and Narungga origins and an activist against racism in Australia, performed a celebratory goal dance. It was a war dance, which triggered great controversy within Australian society. More precisely, it led to a wave of booing in person and digital harassing on social media—part of what Australian media named the "booing saga" against Goodes.[17] In that same year, not long after these events, Goodes retired from football and, in 2016, he deleted his Twitter account. This case is taken up by Ariadna Matamoros-Fernández to investigate what she calls "platformed racism," a term with double meaning: "It (1) evokes platforms as tools for amplifying and manufacturing racist discourse both by means of users' appropriations of their affordances and through their design and algorithmic shaping of sociability and (2) suggests a mode of governance that might be harmful for some communities, embodied in platforms' vague policies, their moderation of content and their often arbitrary enforcement of rules" (2017, 931).

Using an issue mapping approach around the Goodes controversy on Twitter, Facebook, and YouTube, the author tracks the actors, issues, and objects involved. The three platforms, following Matamoros-Fernández, have very different moderation policies when it comes to hate speech—disguised, in many cases, under the form of humor. The author analyzes a corpus of 2,174 tweets with images, 405 Facebook links, and 529 YouTube links shared on Twitter between May 29 and September 16, 2015. In addition, to determine the role of algorithms in ranking contents, Matamoros-Fernández also creates ad hoc profiles on Facebook and YouTube and analyzes the first twenty-five pages suggested by the respective algorithms after deliberately liking pages associated with booing Goodes.

The author finds similarities and differences in how phenomena unfolded across the three platforms. First, on Twitter "attacks

towards Goodes were articulated by means of sharing memes"
(Matamoros-Fernández 2017, 938), which were covered by users who
used "sensitive media" filters. Second, on Facebook "humour tended
to concentrate in compounded spaces, like meme pages, or in comments" (2017, 938). Third, on YouTube "parody was also located in the
comment space rather than being mediated through videos uploaded
specifically to make fun of Goodes" (Matamoros-Fernández 2017, 938).
Regarding recommendations from algorithms, Matamoros-Fernández
notes that recommendation systems reinforced racist content: "[B]y
liking and watching racist content directed to Adam Goodes on Facebook and YouTube, the platforms' recommendation algorithms generated similar content about controversial humor and the opinions
of Australian public figures known for their racist remarks towards
Aboriginal people" (Matamoros-Fernández 2017, 939).

An illustration of qualitative methodologies is Loes Bogers, Sabine
Niederer, Federica Bardelli, and Carlo De Gaetano's (2020) examination of the depiction of motherhood, in particular representations
of pregnancy across the Web and six platforms: Facebook, Instagram, Pinterest, Reddit, Tumblr, and Twitter. The authors resort to
two visualization methods, which they call image grids and compos-
ite images (Bogers et al. 2020, 1043). These methods help reflect the
ways in which platform algorithms order the content they classify
as more relevant, which, in turn, allows observing similarities and
differences among the objects of comparison. The goal of Bogers and
colleagues is to reveal and deconstruct gender stereotypes and biases
that operate on representations of motherhood and that are generated at the intersection of the practices of users and the algorithms
of platforms.

Informed by critical feminist perspectives, the authors discuss the
uniqueness of each platform in terms of "platform-specific *vernaculars*" (Bogers et al. 2020, 1038; emphasis in the original)—defined by
Gibbs et al. (2015) as "the unique combination of styles, grammars,
and logics" (257)—and find significant differences. After an analysis

of the 200 most relevant images associated with the keywords "pregnant" or "pregnancy" on each selected platform, the authors find, among other things, that "Facebook, for example, highlights the heteronormative family unit that is celebrated (and sometimes mocked), while Twitter offers a discourse of information sharing, more pluriform relations and support between women" (Bogers et al. 2020, 1054). Ultimately, the case "confirms the overall lack of various pregnant corporealities . . . and online absence of non-heteronormative ways of doing pregnancy" (Bogers et al. 2020, 1056). The authors argue that platform vernaculars play a role in the distribution of visibility and invisibility of different representations of pregnancy.

This section focused on two alternative methodologies used in cross-platform studies. Matamoros-Fernández (2017) showed the divergent ways in which users engaged platforms' affordances to spread hateful content, reinforced by platforms' algorithms; Bogers et al. (2020) demonstrated that various platforms, although sharing a reification of certain forms of pregnancy, also exhibited significant differences across them. In both cases, particularities regarding a single platform, and similarities and differences with others, would have remained opaque without a comparative approach.

Interpretations

There are two central interpretations that cut across the findings generated by cross-platform research. On the one hand, some scholars posit that platforms have different affordances capable of producing divergent effects (Papacharissi 2009; Utz, Muscanell, and Khalid 2015; Duguay 2016; French and Bazarova 2017; Shane-Simpson et al. 2018). On the other hand, other scholars show that although some functionalities are similar across platforms, user practices contribute to producing divergent modes of appropriation (Larsson 2015; Karapanos, Teixeira, and Gouveia 2016; Zhao, Lampe, and Ellison

2016; Boczkowski, Matassi, and Mitchelstein 2018; Scolere, Pruch- niewska, and Duffy 2018). These interpretive frames are descendants of older debates in the study of technology and society often seen through the prism of technological determinism versus that of social construction (Bijker, Hughes, and Pinch 1987; Grint and Woolgar 1992; Kling 1992; Marx and Smith 1994; Boczkowski and Lievrouw 2008).

A study by Matthew Pittman and Brandon Reich (2016) illus- trates the first option. Interested in examining the impact of social media on feelings of loneliness in young adults, the authors com- pare five platforms: Instagram, Snapchat, Twitter, Yik Yak, and Face- book. Focusing on "which aspects of mediated communication confer experiential aspects that might lead to a genuine social pres- ence of immediacy *and* intimacy" (2016, 157; emphasis in the origi- nal), Pittman and Reich propose a typology of platforms associated with whether their functionalities privilege images or text. In their words, "by focusing on the primary modality of each platform— text or image/video—we might begin to understand how they each mitigate or exacerbate loneliness" (Pittman and Reich 2016, 156).

According to their categorization, whereas Instagram and Snap- chat fall on the image-based platform type, Twitter and Yik Yak fall on the text-based one and, having both image and text modali- ties, Facebook sits in the middle. Based on Shyam Sundar's (2008) MAIN model (Modality, Agency, Interactivity, and Navigability), Pittman and Reich explain that images have a stronger capacity in emulating social presence, which might be associated with a decrease in feelings of loneliness. In order to test this, they conduct a mixed-methods, quasi-experimental survey with 253 college stu- dents. They find that "image-based platforms such as Snapchat and Instagram confer to their users a significant decrease in self-reported loneliness. . . . This ability to mitigate an undesirable psychological state and induce positives ones may be due to the ability of images to facilitate social presence (Sundar 2008), or the sense that one is communicating with an actual person instead of an object" (Pittman and Reich 2016, 164).

A study by Pablo J. Boczkowski, Mora Matassi, and Eugenia Mitchelstein (2018) serves to illustrate the second type of interpretation. It seeks to understand how young adults in Argentina manage five social media platforms—Facebook, Instagram, Snapchat, Twitter, and WhatsApp—in their everyday lives. They ask two questions: "First, what are the dominant constellations of meaning constructed around social media among young people in Argentina? Second, how do different constellations relate to the practices usually enacted on one particular platform in relation to other normally-accessed platforms?" (Boczkowski, Matassi, and Mitchelstein 2018, 246).

The authors find that even though these platforms share several key affordances, a distinct constellation of meaning emerges for each platform: "WhatsApp is a multifaceted communication domain; Facebook is a space for displaying the socially acceptable self; Instagram is an environment for stylized self-presentation; Twitter is a venue for information and informality; and Snapchat is a place for spontaneous and ludic connections" (Boczkowski, Matassi, and Mitchelstein 2018, 245). Furthermore, they contend that "people use one platform in ways related to how they use the others. Second, users' perceptions and sense-making of each platform often include recursive references to other social media options" (Boczkowski, Matassi, and Mitchelstein 2018, 255).

This section presented two possible ways of interpreting the results emerging from cross-platform accounts. In both studies, it is evident that the comparative gaze was beneficial to elicit various phenomena tied to social media uptake and its broader implications. By contrasting two different types of platforms, Pittman and Reich (2016) emphasized how certain functionalities differently affected feelings of loneliness in young users. The research by Boczkowski,

Matassi, and Mitchelstein (2018) showed how platforms sharing similar functionalities could nonetheless be associated with significantly diverse social media cultures shaped by users.

Conclusions

We opened this chapter with two radically different vignettes. The first centered on the concern of Twitter users about the apparent homogeneity and lack of originality across platforms that had become assumed after the announcement of Twitter Fleets. While seemingly banal, it portrayed a picture of high levels of similarity in the social media landscape, an issue that was not well-received by users, who created and circulated memes making fun of it. The second focused on Western political leaders' requests for social media companies to make visible and harmonize their moderation policies in order to prevent hateful and violent content in the wake of the Christchurch mosque shootings. This event, which represented a tragic moment in New Zealand's public life that affected and continues to affect the country's collective memory, and that also had ripple effects across the world, brought to light the heterogeneity across platforms that exists regarding their policies and practices of content moderation. From the banal to the tragic, both vignettes converge to signal the value of adopting a cross-platform lens to understand the appropriation of social media and their cultural and political consequences. Users, from ordinary citizens to heads of state, regularly engage in comparisons when they make sense of the role of platforms in everyday life, and so should scholars who study social media.

To this end, in this chapter we engaged with studies that examined a multiplicity of practices of use that included comparisons of eleven platforms: Facebook, Instagram, Pinterest, Reddit, Snapchat, Tumblr, Twitter, Vine, WhatsApp, Yik Yak, and YouTube. A cross-platform comparative stance enables the analyst to unpack the monolithic notion of social media that has arisen from single-platform studies which assume that what happens on one platform might apply to the social media landscape as a whole, thus implying notions of a homogeneous and undifferentiated unity (Tufekci 2014; Bode and Vraga 2018; DeVito, Walker, and Birnholtz 2018). In contrast, the image that emerges from cross-platform accounts is that social media should not be treated as an a priori unified object of inquiry (Hall et al. 2018; Hargittai 2020; Yarchi, Baden, and Kligler-Vilenchik 2020). Platforms are by definition plural, and what might differentiate and/or unite them should be discovered instead of being taken for granted.

As we stated in chapter 1, these ideas draw from several theoretical developments in the field of communication studies that have fostered relational and ecological accounts of media use, such as niche theory, repertoires, and polymedia.

The theory of the niche (Dimmick 2003; Dimmick, Feaster, and Ramírez Jr. 2011; Ha and Fang 2012) explores how different media survive and grow in a changing environment. It was originally formulated to understand the competition between old and new media: "A new medium will compete with established media for consumer satisfaction, consumer time, and advertising dollars. If competition does exist, then the consequence for the older media consists of exclusion or replacement, or displacement, wherein the new medium takes over some of the roles played by the older medium" (Dimmick, Chen, and Li 2004, 22).

Although this focus brings us back to dynamics cutting across traditional and social media that were addressed in the previous chapter, niche theory can also be applied to explain relationships across platforms. The impetus for the imitation of the stories functionality originally developed by Snapchat and subsequently replicated by Instagram, Facebook, WhatsApp, and Twitter in their attempt to remain current with their user base, especially its youth segment, and fend off migration to Snapchat, provides a clear illustration of the potential of niche theory for comparative cross-platform accounts.

While niche theory stresses market dynamics, scholarship about media repertoires (Hasebrink and Popp 2006; Taneja et al. 2012; Hasebrink and Hepp 2017; Swart, Peters, and Broersma 2017) examines phenomena from users' points of view. As we mentioned in chapter 1, this theoretical lens was originally conceived to understand decision making in the face of significant increases in programming options during the transition from terrestrial to cable television. In doing so, it allows us to understand how users assemble their own mix of content from multiple sources, thus forming a repertoire. In the words of Taneja et al. (2012), "Studies have consistently found that users do not divide their time consuming all available media (e.g., Heeter, 1985). They instead create subsets of all available options and consume content from this smaller set. These subsets are referred to as repertoires. Almost all early studies on repertoires were focused on repertoire formation in television viewing. These consistently found that, on average, viewers watched a fraction of television channels received by their household. Subsequent studies have expanded the concept of repertoires beyond television viewing, to interpret consumption patterns across multiple media" (953).

A number of social media studies conducted over the last decade have shown that the ways in which users appropriate platforms indicates that they do so building repertoires; they use multiple platforms but not all of them, for different purposes, sometimes strategically and others ritualistically, thus creating their own social media repertoires (Zhao, Lampe, and Ellison 2016; DeVito, Walker, and Birnholtz 2018; Boczkowski 2021).

Also introduced in chapter 1, the theory of polymedia, developed by Mirca Madianou and Daniel Miller (2012) and subsequently adopted in a number of social media studies (Renninger 2015; Madianou 2015, 2016; Boczkowski, Matassi, and Mitchelstein 2018; Tandoc Jr., Lou, and Min 2019), adds a cultural and relational sensibility to the market-centric approach of niches and the user-centric view of repertoires. Madianou and Miller argue that

> polymedia is an emerging environment of communicative opportunities that functions as an "integrated structure" within which each individual medium is defined in relational terms in the context of all other media. In conditions of polymedia the emphasis shifts from a focus on the qualities of each particular medium as a discrete technology, to an understanding of new media as an environment of affordances. As a consequence the primary concern shifts from an emphasis on the constraints imposed by each medium (often cost-related, but also shaped by specific qualities) to an emphasis upon the social and emotional consequences of choosing between those different media. (2013, 170)

The theory of polymedia refers to interpersonal relationships and the focus shifts to examining the emotional and social factors that shape how users decide to integrate different platforms as part of their ongoing relationships, enacted within particular local contexts.

The aforementioned three theoretical frameworks provide us with a useful toolkit to tackle comparative work involving multiple platforms. Cutting across all of them is the idea that the unit of analysis of comparative research on social media platforms can be considered relationally (Hasebrink and Popp 2006). Put differently, in cross-platform comparative research the unit of analysis shifts from what happens within a given platform into the relationships across two or more of them. This, in turn, enables the analyst to figure out what is unique to a particular platform and what is shared with others. If, as Lisa Gitelman's (2006) assertion that traditional media *are* also applies to their social media counterparts, then the pluralization of platform practices necessitates acknowledging their differences as something intrinsic to them. This does not mean that these differences will always matter or, if they do, that they will always matter in similar ways. But it means that from a comparative social media studies standpoint, scholarly accounts should inquire if, when, how, and why these cross-platform differences make a difference.

A by-product of this pluralization of our understanding of platforms is that it can act as an epistemic antidote to the deterministic tendencies that have dominated academic and popular discourses on social media in recent years. The Brexit referendum in the United Kingdom and the elections of Donald Trump and Jair Bolsonaro as presidents of the United States and Brazil, respectively—among other recent political events in the world—revived explanations based on hypodermic needle tropes of social media altering the will of the people and the integrity of democratic processes. These explanations often elided any differences across platforms and overlooked the agency of their users. By design, cross-platform perspectives move these differences to the foreground. Because many of the key affordances that are credited with the presumed negative outcomes of social media on society are shared by the main platforms, variance in use across platforms cannot be attributed to the affordances themselves. This, in turn, invites us to switch the attention to the agency of users and the various kinds of possible interactions with the structuring power of technology.

Another important antidote to the deterministic tenor of currently dominant discourses on social media has to do with understanding their histories and how they relate to the histories of other media. It is to this matter that we turn next.

II Pathways

II. Pathways

5
Histories

Introduction

Groundhog Day

On September 9, 2020, Netflix releases *The Social Dilemma*, one of the most publicly discussed films of that year. It is a fictionalized documentary about potential risks of using social media, with testimonials from former high-ranking workers at technology companies in Silicon Valley. Most interviewees have been involved firsthand in the creation of products such as Gmail, Facebook, and Instagram. The protagonist is Tristan Harris, an expert on the prevention of the psychological, social, and political harms presumably associated with contemporary information and communication technologies. Harris's myth of origin has to do with the disenchantment he experienced as a design ethicist at Google. Such disenchantment ultimately led him to become director of the nonprofit Center for Humane Technology. Since then, the Center has been dedicated to conducting awareness campaigns to alert society about the dangers contained in the design of supposedly addictive mobile technologies and social media.

The central message of *The Social Dilemma* is that technology companies deliberately profit from the attention of their users and manipulate it, with potentially dangerous political implications. To make the case, the script of the film constructs, often implicitly, a model of the user as someone devoid of agency and self-reflexivity, and therefore amenable to having their thoughts and actions being directed by others. Social media are correspondingly imagined as the equivalent of an addictive and unhealthy drug that robs users of their autonomy, will, time, and relationships—and is ultimately able to destroy democratic regimes.

Against the backdrop of the alleged novelty of these technological innovations, it is worth recalling Ellen Wartella and Byron Reeves's (1985) indication that the moral panics associated with the use of communication and information technologies show a recurrent pattern throughout modern history. These authors quote May Seagoe who in 1951 argued that "Whenever there is a new social invention, there is a feeling of strangeness and a distrust of the new until it becomes familiar" (143). Amy Orben (2020) calls this "the Sisyphean cycle of technology panics." Like Sisyphus in his myth, who lifts a giant boulder every day to a peak only to have it fall again, individuals and organizations have repeatedly produced dystopic discourses about the harms that media technologies can bring to individuals and collectives—from the codex to the radio to movies to the internet. Such discourses, which are sustained by what Leo Marx and Merritt Roe Smith (1994) call "the 'hard' end of the spectrum" (xii) of technological determinism, always seek to protect those considered weaker from the dangerous seduction of new devices and their potentially deleterious effects. Along the way, these discourses encourage governments to promote applied research on media effects and scholars to propose linear models of causes and effects. The result is a state of affairs in which current models fail to build on prior scholarship and research yields inconclusive findings, which are also forgotten by the time the next major innovation comes around. At that point, fear emerges again and the cycle starts from scratch.

There is one scene in *The Social Dilemma* that illustrates this pattern with particular resonance for those familiar with the social and historical studies of technology. Tristan Harris looks into the camera and explains that to understand the unique seriousness of social media's effects relative to those of other technologies of the past, it is enough to know that "no one got upset when bicycles showed up." That assertion conveniently ignores that one of the foundational academic works in scholarship on the social construction of technology (Pinch and Bijker 1984) recounts the various tensions across multiple social groups that arose in relation to the development of bicycles in the late nineteenth century. For a moment, it seems as if viewers are watching not Tristan Harris but Phil Connors, the character played by Bill Murray in Harold Ramis's iconic 1993 film *Groundhog Day*.

The Multiple Genealogies of Platforms

The Social Network, released a decade earlier than *The Social Dilemma*, tells a story about Facebook's origins. The film suggests that one key antecedent of Facebook was the website FaceMash. As stated in Mark Zuckerberg's deposition before the US Congress in 2018, FaceMash "was a prank website that I launched in college, in my dorm room, before I've started Facebook."[1] That website, which was kept live for a few hours, proposed a game based on the comparison of women's images taken from Harvard University's directory. *The Crimson* called it the "Harvard version of the Am I Hot or Not? website,"[2] referring to a site created in 2000 by two University of California at Berkeley engineering graduates that ranked people's beauty on a scale of 1 to 10. *The Social Network* suggests a genealogical line connecting Face-Mash and Facebook.

Shortly after Facebook launched in the United States, Niconico (2006) was created in Japan and Sina Weibo (2009) in China. In each of these cases there appear to be genealogies that diverge from that of Facebook. Jack McLelland, Haiqing Yu, and Gerard Goggin (2018, 53) remind us that "we are moving away from a time when discussions about the Internet and the effects of its myriad applications can be discussed or judged from an exclusively North American (or even wider Anglophone) perspective." One potentially generative angle into this can be found in the docuseries *High Score*, also released in 2020 like *The Social Dilemma*, which narrates the historical development of video games. It shows how many of the ideas and aesthetics of the most popular games of the second half of the twentieth century and the first two decades of the twenty-first century emerged in no small measure within the creative scene of Japan in the 1970s and 1980s. The docuseries begins by focusing on the figure of Tomohiro Nishikado, creator of *Space Invaders*. This game diffused from Japan to the United States. It became a commercial hit among American kids and teens, and subsequently in many other parts of the world. It was a pioneer in the "shoot 'em up" genre and inaugurated the use of "high scores" to catch the attention of players.

Currently, the video game industry provides one of the great sources of content and lifestyles on social media platforms such as Twitch (Taylor 2018; Gray 2020)—where the goal is to play online and watch others play, as we noted in chapter 3 regarding the streamer-turned-political commentator Hasan Piker. Perhaps relatedly, the first algorithmic idea that gives rise to the Facebook prototype, according to *The Social Network*, has to do with playing a misogynistic game online. Could Facebook exist without a culture of gaming? Can a platform invented in Cambridge, Massachusetts, at the dawn of the twenty-first century, have roots in the Tokyo creative scene of the 1970s? Do platforms have multiple genealogies? More generally, can we understand the present of social media without comprehending their past?

Why Historical Comparisons Matter

The role of history in the nascent scholarship on social media has been decidedly relegated in favor of a present-day focus. That is, the objects of inquiry examined in the present tend to be implicitly naturalized as ahistorical phenomena and therefore their histories are left out of most scholarly analyses. However, the two vignettes presented earlier show that a historical look at platforms can shed light on their evolution over time, their modes of use, and their social and political consequences. This, in turn, can help illuminate continuities and discontinuities that are fundamental to a better understanding of what might be unique about the present. For it is precisely by observing the historical links between phenomena across different points in time that it is possible to identify areas of discontinuity. These areas are ultimately what indicate transformations and novelty. In the words of Ben Peters (2009, 15): "Alone, neither continuity nor change approaches to media history are fully satisfactory. However, viewed together, they complement one another: the historian's eye for contingent change can lead to a fuller understanding of the contemporary relevance of media; so too can new media scholars engage the present more forcefully with historiographical cautions in mind."

The first vignette, on the cyclical nature of technology panics about media and communication technologies, allows us to place the dominant apocalyptic tone of contemporary conversations about social media within a long tradition of similar dystopic discourses. In other words, from a historical standpoint there is little novelty in *The Social Dilemma* and the related narratives that circulate in contemporary society. The second vignette, on the multiple genealogies of platforms, provides one concrete illustration of the many influences of the past in the present and therefore opens up the possibility that platforms can have rich and complicated histories. One common account of the origins of Facebook centers in the United States during the dawn of the twenty-first century and magnifies through the connection with FaceMash the misogynistic biases of algorithmic cultures examined by Safiya Noble (2018) in the case of search engines. Another alternative yet complementary account takes us back to Japan several decades earlier and highlights the role of transatlantic flows of gaming cultures over time.

In both vignettes, the historical perspective makes it possible to identify the influence of the past in the present and therefore also what might be novel about a platform in particular and social media in general. However, the dominant approaches in social media studies have constructed—sometimes by denotation although more often by connotation—a present that lacks a past and therefore hinders the ability of the analysts to assess the meaning and implications of the phenomena under study. To counteract this tendency, in this chapter we explore historical pathways that enrich the comparative gaze in social media scholarship. For although comparing contemporary phenomena can teach us many things, adding a historical layer allows us to challenge the inevitability of conceptual assumptions and interpretations of empirical findings that have been often baked into present-day biases.

History in Cross-National and Regional Comparisons

Antecedents

A study by Marc Steinberg (2020) that examines the evolution of LINE in Japan provides a fruitful antecedent to show the importance of history in social media matters. Steinberg argues that the emergence of LINE is inextricably tied to a local culture of mobile connectivity marked by i-mode and dating back to the late 1990s in Japan. More precisely, LINE's historical evolution has been shaped by a visual culture represented by the large sticker collections that the platform hosts. According to the author, LINE draws "on emoji and deco-mail proto-stickers pioneered by i-mode, and on the wider character-centric visual culture of manga, anime, and games, including manga's complex grammar of semiotic signs used to denote emotions" (Steinberg 2020, 5). Furthermore, Steinberg argues that it is from this platform that stickers began to be imported into other platforms such as Facebook and WeChat.

These visual artifacts are sold by amateur producers who are part of a "Creator's Market" within the LINE platform. While this "entrepreneurialization of the subject" (Steinberg 2020, 7) could be read as a descendant of the neoliberal culture of Silicon Valley, Steinberg shows the importance of historicizing the phenomenon to properly appreciate its meaning within broader patterns of Japanese labor market trends: "This is where I take issue with the platform presentist and American-centric framings of contemporary labor conditions. . . . [The entrepreneurialization of the subject] is also, and in equally large part, an extension of the progressive increase in contingent work underway, at least, since the deregulation of the labor market in Japan in the 1990s . . . , and present even during the height of Japan's economic growth in the 1970s and 1980s in the form of automobile and electronics factories' subcontracted, precarious labor" (Steinberg 2020, 7).

Another study that demonstrates the power of historical analysis is the account by Luolin Zhao and Nicholas John (2020) of "sharing" in Chinese social media. The notion of sharing has become a central element of platform use. According to Nicholas John's (2013, 2017) previous work, in the West it has been tied to three semantic fields: therapy, computing, and economics. In China, however, the same concept has been linked to a double translation: *fenxiang*, which has to do with dividing and distributing, and *gongxiang*, which implies the action of enjoying together. Zhao and John (2020) contend that a historical sensibility is essential to understanding the culturally situated enactment of both terms. Thus, they focus on their long evolution: from the Qing Dynasty (1636), in the case of *fenxiang*, and from the Han Dynasty (206 BC), in that of *gongxiang*. They explain that "while *fenxiang* has gradually transformed from dividing and distributing into an act of communication with interpersonal connotations, *gongxiang*'s newer meanings lie in the technical realm, while conveying and promoting the value of sharing and harmony in a higher societal sense" (Zhao and John 2020, 7).

Zhao and John complement this focus on the *long durée* of linguistic evolution with another one about transformations in the recent history of individuality in China. They argue that what they call a "divided self" is currently taking shape, where notions of appreciation of individualism coexist with a high respect for state authority. Therefore, "In the context of Western social media, 'sharing' (or at least an ideal type of 'sharing') appeals to people who authentically communicate their true core selves, and, according to SNSs [social networking sites], is a practice that will bring about better interpersonal understanding. In the context of Chinese social media, *fenxiang* appeals to people who wish to communicate within a reciprocal relationship while expressing themselves in a risk-free, altruistic manner while for the SNSs, *gongxiang*, the state attained by *fenxiang*, will bring about societal harmony, in keeping with the state's objectives" (Zhao and John 2020, 14).

These two studies illustrate how a historical view helps illuminate both shared and unique patterns in the appropriation of platforms around the world. Despite the contributions enabled by the historical work, both studies seem to engage in cross-national and/or regional comparisons in an ad hoc fashion. In the next section we chart some possible future directions of research that compare the temporal evolution of social media practices in two or more national or regional contexts in a programmatic fashion.

Future Developments
We outline two possible lines of future work. The first one has an institutional sensibility and centers on examining how different national contexts shape divergent trajectories of the same platform. The recent history of the incorporation of payment functionalities to WhatsApp provides an interesting example for this approach. WhatsApp Pay debuted in Brazil on June 15, 2020, but eight days later it was blocked due to antitrust concerns. This triggered a nine-month-long dispute between the Brazilian Central Bank and WhatsApp about the

potentially negative impact of the app on the country's local banks and financial technology companies. The government requested the company to adapt WhatsApp Pay to PIX, the Central Bank's own instant digital payment system. During the press conference in which PIX was originally launched, Roberto Campos Neto, then President of the Central Bank of Brazil, claimed: "WhatsApp will start doing P2P soon. I have talked a lot with their CEO, we are making good progress. He has told me that the process (with us) *was faster than in other countries*"[3] (emphasis is ours). The pay functionality finally was approved on March 30, 2021.

Why was this process "faster than in other countries"? To answer this question, it helps to examine what happened with the implementation of WhatsApp Pay in India. In this country the incorporation of this functionality took place in 2020, following a protracted four-year negotiation process that presumably started after Prime Minister Narendra Modi's 2016 attempt to demonetize the Indian economy. This measure, which consisted in severely restricting the circulation of cash in a largely informal economy, was part of Modi's larger "Digital India" plan launched in 2015 and publicly backed by Mark Zuckerberg.[4] Despite the alignment between the company and the government, a series of regulatory disputes and technical adjustments greatly slowed down the incorporation of the pay functionality into the app. The Indian government approved the use of WhatsApp Pay in November 2020, when it conformed to India's Unified Payments Interface.[5] In the words of Ravi Shankar Prasad, then minister of Information Technology, Law, and Justice, "India is the world's largest open Internet society and the Government welcomes social media companies to operate in India, do business and also earn profits. However, they will have to be accountable to the Constitution and laws of India" (Singh 2021).[6]

The contrast in the implementation of WhatsApp Pay between Brazil and India raises a key question at the intersection of comparative historical work: Was the process faster in Brazil than in India because of (a) a learning effect within the company, (b) a difference between the political, regulatory, and/or technological systems in the two countries, (c) a combination of both, or (d) none of the above? Answering this question necessitates a cross-national comparative historical perspective, one that is attentive to the consequences of both the passage of time and the differences across countries. Furthermore, this perspective could help illuminate the dynamics of expansion of platforms from one country to another in the context of a market that is highly concentrated in the hands of a few players and where demands for national and international regulation seem to be increasing—consider, for instance, the vignette presented in chapter 4 concerning New Zealand and the Christchurch attack.

The second avenue for future work that we propose, from a cultural and political economy perspective, would center on de-westernizing social media genealogies (Curran and Park 2000). One illustrative example could be that of deconstructing the constitutive ties between social media based in the United States and the libertarian value system, entrepreneurship ethos, and close connection to elite university research of Silicon Valley (Lécuyer 2006; Turner 2006; Streeter 2011; Marwick 2018; Meehan and Turner 2021). This would imply revising two separate yet interrelated issues: the history of Silicon Valley and its relationships to social media companies, and its global status as both role model and default imaginary for other locations of social media production in the world. The former topic would entail shedding light on the contingent economic and social decisions that over time turned what was essentially a farmland into "a regional network-based industrial system that promotes collective learning and flexible adjustment among specialist producers of a complex of related technologies" (Saxenian 1996, 2). In other words, Silicon Valley as we know it today is a recent historical achievement. Uncovering that history and unpacking the processes that guided it would allow scholarship on social media to challenge assumptions about how and why platforms have been designed, built, marketed, and distributed in certain ways and not others.

Restoring the historical contingencies that led Silicon Valley to acquire its contemporary status and interrogating the connections of this history to social media would be tied to interrogating its global implications. According to Marwick (2018): "[D]espite its excesses, Silicon Valley functions as a global imaginary: it models what is considered a superior type of wealth-generating innovation for other places eager to replicate its success. Thus, we must take it seriously as attempts are made world-wide to replicate its practices" (314).

A cross-national historical comparative perspective would problematize attempts to conceive of Silicon Valley as a benchmark for other regions in the contemporary sociotechnical imaginary (Jasanoff and Kim 2015). This tendency is reflected, for instance, in the ways in which the anglophone news media often characterize Zhongguancun as the "the Silicon Valley of China," or Bangalore as "the Silicon Valley of India." This tendency not only unreflexively exports American models into other parts of the world but also obscures what might be unique about what goes on in different locales. Thus, attending to the development of alternative platforms such as WeChat in China—created with the strong presence of a large bureaucracy like Tencent and within the environment of a state-controlled economy—could help denaturalize the dominant Silicon Valley global narrative. Historical cross-national research could contribute to show both the limits of this narrative and its associated sociotechnical imaginary, as well as illuminate how the specific histories of different locations around the world contribute to divergent trajectories of social media production and use.

Historical Cross-Media Comparisons

Antecedents

How are we to think about the relationships between traditional and social media in historical ways? One powerful illustration can be found in the work of Lee Humphreys, who in her book *The Qualified*

Self: Social Media and the Accounting of Everyday Life (2018) and in papers with colleagues, has sought to establish the historical roots of contemporary social media practices in much older ones, such as personal diaries. For instance, Humphreys et al. (2013) explain that in the eighteenth and nineteenth centuries the personal diary genre had a semipublic trait characterized by the reflection of daily life in a brief format and, in general, was devoid of depictions of emotional states. Contrary to what is usually imagined about personal diaries as necessarily private, the authors show how they in many cases traveled from one place to another and were shared with loved ones and nearby communities to give an account of the events of one's own life. Comparing the content of personal diaries with a sample of tweets from the year 2008, Humphreys et al. (2013) find that the actors mentioned in most cases are the authors of the tweet and, to a lesser extent, other person(s), thus resulting in a mix quite like that of social diaries centuries ago. So "rather than condemn the accounting and reflecting practices on Twitter as narcissistic (Sarnow 2009), by placing them into a longer discussion of media and communication we can begin to understand Twitter's popularity. While there are important differences . . . the similarities to historical diaries suggest long-standing social needs to account, reflect, communicate, and share with others" (Humphreys et al. 2013, 428).

Another fruitful antecedent of historical accounts of cross-media comparisons is provided by Bridget Kies (2021), who analyzes the case of the *Jimmy Kimmel Live!* television show. She focuses on its segment "Celebrities Read Mean Tweets," presented by Kimmel as an instance of "encounter" between celebrities and social media audiences, in which "a celebrity reads the tweet from a phone while the tweet and Twitter user's handle is displayed on screen" (Kies 2021, 517). The author's goal is to ultimately understand the changing interactive dynamics between television and social media. She thus traces the historical evolution of "celebrity roasts" as a form of media event. This format emerged from the New York's Friars Club back in the early twentieth century and was then televised in 1968 with NBC's *Kraft Music Hall*. Subsequently it started to be produced as a stand-alone format on Comedy Central in the early twenty-first century. Kies (2021) finds that "As televised roasts move further from their origins to more closely resemble the bullying and trolling found on social media, the use of mean tweets on late-night television segments like 'Celebrities Read Mean Tweets' becomes a contemporary remediation of the celebrity roast. 'Celebrities Read Mean Tweets' not only finds its source material on Twitter but remediates it as television" (524).

The research by Humphreys et al. (2013) and Kies (2021) illustrates the power of historicizing the ties between traditional and social media. In the next subsection we build on their contributions to continue developing building blocks of a comparative cross-media agenda that is attentive to historical dynamics.

Future Developments

We propose two potential areas of work to further the historical dimension of cross-media dynamics. The first focuses on the influence of traditional media in the emergence and unfolding of different social media platforms. The second centers on the coevolution of traditional and social media.

There are at least two ways in which we can ascertain how traditional media formations have shaped platforms: genre conventions and defining features. Regarding genre conventions, for instance, research discussed in chapter 3 has highlighted the role of reality television in preparing the ground for the confessional style that has marked the presentation of the self on Facebook. That is, many of the discursive resources that have been common for information presentation and commentary on that platform have a strong connection with similar resources that were first popularized in reality television and the social uses of webcams in the 1990s and in blogs in the 2000s (Holmes and Jermyn 2004; Koskela 2004; Kraidy 2009;

Siles 2017; Psarras 2020). Furthermore, the centrality of immediacy, the prevalence of sound bites, and the role of strong opinions that have been the hallmark of Twitter have a direct antecedent in the contemporary evolution of journalistic conventions, in both print and broadcast media. Thus, it is unsurprising that journalists themselves have gravitated to Twitter as their platform of choice both for gathering and disseminating information (Hermida 2010; Lasorsa, Lewis, and Holton 2012; Paulussen and Harder 2014).

Moreover, the aestheticized presentation of the self that characterizes Instagram has strong ties to the celebrity system that has been part and parcel of mediatized entertainment since the dawn of mass media and that has intensified in recent decades with dedicated programming in cable television and the tabloidization of print, broadcast, and digital journalism (Douglas and McDonnell 2019). The very notion of influencer, one of the supposedly novel aspects of social media, cannot be fully understood in its continuities and discontinuities without placing it in a long lineage of practices of mediatized parasocial interactions with celebrities and the system that manufactures and sustains them (Marwick 2013; Duffy 2017; Christin and Lewis 2021; Craig, Lin, and Cunningham 2021). Finally, the carnivalesque genre that permeates some of the most recent platforms such as Snapchat and TikTok has important antecedents in the carnivals, fairs, and magic shows from the nineteenth and twentieth century (Hill 2011; Jones 2017)—which also continue to this date, with varying degrees of popularity. The prevalence of visual tricks, costumes, and masks, and seemingly more spontaneous and less inhibited behavior that is expected in the use of these platforms, has an uncanny yet seldom explored resemblance to those prior mediated ways of staging experiences of enjoyment and awe.

Cutting across these different ties between traditional and social media is how much the genre conventions of the latter have been shaped by those of the former media. Also at play are specific areas of discontinuity in which the present differs from the past. In both cases, a historical gaze is key to expanding and enriching our knowledge of what is and is not new about social media platforms in comparison to traditional media counterparts.

Concerning defining features, and going back to the topic of one of the studies discussed earlier, one of the most central elements of social media has been the act of sharing, which has critical historical antecedents in traditional media. First, several scholars have pointed out the extent to which sharing has been central to the emergence and development of earlier media and communication technologies (John 2013, 2017; Hermida 2014; Hartley 2018). For instance, the first newspaper published in what would eventually become the United States, *Publick Occurrences, Both Forreign and Domestick*, printed its first and only issue in Boston on September 25, 1690. It had four pages and the publisher, Benjamin Harris, only printed news in the first three, leaving the fourth page blank so that readers could annotate their news before passing along the issue to other members in their community (Emery and Emery 1978). Furthermore, as noted in chapter 1, the research by Douglas (1989) has shown that amateurs played a decisive role in the transition of the radio from a point-to-point technology to a mass medium. Their desire to share their favorite aural content and communicate with fellow amateurs was a critical aspect in the historical development of radio as we have come to know.

Moreover, Fischer (1992) has demonstrated that the telephone, originally designed and marketed as a technology to support business communication, became a central element in the communication infrastructure of everyday life of the twentieth century due to the unforeseen development of regular users taking up the artifact during nonwork hours for noncommercial purposes. A common denominator across these histories of newspapers, radio, and landline telephony is the agency of users appropriating new technologies to share what is important to them. Understanding the role of sharing on social media is therefore enriched by establishing the connections with sharing in earlier technologies and also the potential areas of novelty in the case of one or more platforms.

Concerning the coevolution of traditional and social media, for instance, over the past decade news organizations have regularly added social media posts to their repertoire of sourced content (Paulussen and Harder 2014; von Nordheim et al. 2018; Bouvier 2019). It is common to read, for example, articles that curate series of tweets, Instagram posts, or viral TikTok dances to convey a news story. In turn, sources seem to have adapted their social media practices over time to maximize the likelihood of their posts being picked up by journalists. The Kardashian-Jenner vignette presented in chapter 3 illustrates this state of cross-media awareness: televised scenes showcased on platforms and social media posts being subsequently featured on tabloid covers.

One news genre in which this coevolution of traditional and social media has become particularly salient is that of stories concerned with the passing of a public figure. According to Moran Avital (2021, 1,742), "the media have become the main social platform in which public grief is constructed and delivered. By telling stories of death, the media provide the means and opportunity to discuss shared values and moral lessons." Reporters have recently resorted to farewell social media posts as source materials for their stories. This, in turn, seems to have increased the level of self-consciousness of social media users regarding their posts. In the words of Davide Sisto (2020, 181), "every time a famous musician, actor, writer, or sports figure dies, social media users compete to see who can write the most poignant message, or share the most iconic images, video clips, and quotes from that celebrity's career."

Footballer Diego Armando Maradona passed away on November 25, 2020. Within minutes the home pages of news sites around the globe were filled with articles about it, some of which were devoted to the repercussions of the story on social media. The Italian *La Repubblica* published a story titled "Farewell to Diego Maradona, from Pelé to Messi: The messages of condolences on social media,"[7] ESPN's English-language website wrote "Diego Maradona dies at the age of 60: How social media reacted,"[8] and *India Today* reported that "many TV celebs like Sidharth Shukla, Karanvir Bohra and Arjun Bijlani took to social media to pay last respects to football legend Diego Maradona."[9] Did the public figures who posted on social media do so partly with the awareness that news organizations might pick up the content in their stories? If so, did they approach their posts to maximize the likelihood of traditional media exposure, either by themselves or with the help of public relations professionals? Conversely, did reporters strategically canvas platforms in search for quotable posts? Did they even contact potential sources inquiring about their relevant social media activity? Have the editorial practices and genre conventions associated with writing obituaries changed over time in relation to this coevolution? A combination of comparative cross-media and historical sensibilities is helpful to answer these and related questions about the coevolution of traditional and social media. Focusing on only one medium as it evolves or multiple media at one point in time would miss the historical dynamics of mutual shaping (Bijker 1995) that are the heart of the ongoing changes in editorial work and social media practices.

History in Cross-Platform Comparisons

Antecedents

A rich antecedent showing an (albeit implicit) historical perspective on a cross-platform work is D. Bondy Valdovinos Kaye, Xu Chen, and Jing Zeng's study (2021) about what they call the "parallel platformization" of Douyin and TikTok. The authors deem TikTok the international version of Douyin, its Chinese counterpart. Both platforms focus on short videos, a genre with origins in China and with a subsequent successful adoption in the West. According to the authors,

Douyin and TikTok were "developed by the same tech company but deployed in vastly different contexts and have thus far managed to survive as emerging platforms in two opposing but comparable oligopolistic platform ecosystems" (Kaye, Chen, and Zeng 2021, 3). Even though the authors undertake a cross sectional analysis through the app walkthrough method, they make sense of why "the waters of TikTok and Douyin flow from the same source into two highly distinct pools" (Kaye, Chen, and Zeng 2021, 17) since they are guided by a historical sensibility. Thus, they find a key reason why Douyin has developed highly appealing business models for its content producers: "The short video market has had a longer gestation period in China (Su 2019), which is reflected in Douyin's wider variety of options for direct monetization. In addition to virtual gifting, Douyin also includes a "merchandising on behalf" (*daihuo*) feature that embeds icons in live streams that link to products. . . . Merchandising on behalf was pioneered by Chinese online shopping platforms such as Taobao and Mogu" (Kaye, Chen, and Zeng 2021, 14).

Another fruitful antecedent of historical cross-platform work is that of Jessica H. Lu and Catherine Knight Steele's (2019) examination of joy as a resistance strategy by Black users in interactions linking Twitter and Vine. In tracing the long history of Black oral culture as a form of resistance to slavery dating back to eighteenth-century United States, the authors note that "Black rhetorical strategies were rarely employed in isolation from one another. Storytellers used song interwoven with their narratives, and dozens of players alluded to folklore in their verbal play. Likewise, Black users on Twitter are not isolated or limited by singular platform use" (Lu and Steele 2019, 826).

Applying Critical Technocultural Discourse Analysis, Lu and Steele analyze a sample of tweets and vines created around three hashtags aimed at resisting mainstream news portraying negative images of Black children and a proliferation of images of Black death: #carefreeblackkids, #CareFreeBlackKids2k16, and #freeblackchild. They find a cross-platform dynamic that is key to the resistance joy strategy deployed by users in digital spaces: "crossing over and seizing both platforms, brief moments recorded 'live' assert—in multi-sensory fashion—that Black people are fully alive" (2019, 832). Furthermore, Lu and Steele locate the affordances of the platforms as part of a broader, historical repertoire of joy resistance strategies within Black culture: "Because so many of these joyful [Vine] posts incorporate music and dance, they further demonstrate how the affordances of a platform can be made poignant by Black users, in particular. Black users extend traditions of using song as a resilient resistance strategy, especially since Black lyricism and music continue to escape the full understanding of dominant groups" (2019, 832).

The works of Kaye and colleagues (2021) and Lu and Steele (2019) highlight the interpretive gains of a historical gaze in cross-platform accounts. However, this gaze appears to be implemented in an ad hoc fashion rather than in a programmatic one, thus limiting its full potential. Next, we begin developing such a programmatic approach.

Future Developments

We propose two potentially fruitful areas to advance an agenda of comparative historical cross-platform scholarship. The first focuses on recovering the history of abandoned, unsuccessful, or at least marginal platforms, and the second on the coevolution of platforms currently in use by a significant portion of the population. Because social media platforms are relatively recent technological innovations, this agenda is limited to what could be considered recent history. However, even with this limitation we believe a historical sensibility could greatly contribute to more robust cross-platform accounts.

There are at least three reasons that justify a focus on what could be called "dead/dying" platforms (Kluitenberg 2011; Parikka 2012). The first one has to do with accomplishing one of the foundational goals of historical scholarship: presenting a comprehensive account of the past. Complementary to the present-day bias of most research

on social media there is a certain sense of historical erasure of platforms that are no longer in existence—or at least in use by sizable
portions of the public. That is, in addition to the fact that most of the
studies on platforms such as Facebook and Twitter focus on contemporary matters, there is exceedingly limited scholarship on platforms
that are not in use today. It is as if this was a domain of inquiry without a past or as if the past did not matter. Recovering the history of
dead or dying platforms is critical to understanding in general terms
the different ways in which the past might be shaping the present.

The second reason goes into something more specific. The present-
day preference entails by implication the possibility of a success bias
built into research designs. In other words, the practices associated
with the platforms that have concentrated most of the scholarly
attention are in a very basic sense successful because they have managed to persist. However, there are a host of potential alternative
practices in relation to platforms that are either dead or dying that
are not captured by this focus on current practices associated with
platforms in existence. Moreover, it is possible that by not taking into
consideration either these neglected practices or their associated platforms, the explanations provided in relation to the practices studied
regarding platforms in existence might be limited by a sampling on
the dependent variable. In other words, not inquiring into failed
practices and platforms might limit what we are able to know about
successful ones.

The third reason supporting the study of dead or dying platforms
centers on its role in redressing inequalities in social media scholarship at large. This is because even when there is work on these
platforms, it tends to favor some at the expense of others. MySpace
is often cited as one of the greatest commercial ascent-and-descent
cases in the early history of social media. Created in 2003, its period
of splendor in terms of number of users and level of engagement was
between 2005 and 2008, especially in the United States. Several academic studies were devoted to this platform during these years and

in the period immediately afterward, when the ascent of Facebook eclipsed MySpace (Dwyer, Hiltz, and Passerini 2007). Although it is still active and available in up to fourteen languages, its membership has decreased significantly over time. With this decline came a parallel lack of interest among scholars, and there has been very limited work on it since its decline began (Torkjazi, Rejaie, and Willinger 2009). However, despite this neglect there is a bounty of studies on MySpace in comparison to those available about other platforms that had their heyday at one point but are dead or dying nowadays.

For instance, Fotolog is another now-obsolete platform that has received exceedingly limited attention among scholars. Launched a year before MySpace, in 2002, in 2007 it had one of its peaks of success, being listed as one of the twenty most-visited websites in South America. In countries like Argentina and Uruguay, Fotolog was strongly associated with the emergence of an urban tribe—the "floggers."[10] After almost a decade of commercial decline, the platform closed in 2016, only to be resurrected in 2018, but almost anecdotally and from a place of nostalgia (Marcin 2020). Why is the story of MySpace told more often than that of Fotolog? Were their trajectories of rise and fall similar or dissimilar? What makes one remain almost dormant while the other has returned in a nostalgic key but not used massively? Answering these and related questions might provide important insights about patterns of inequality in scholarly attention. This, in turn, should strengthen the historical gaze by broadening the scope of suitable objects of inquiry.

To complement the historical focus on dead or dying platforms, we propose a second pathway centered on the coevolution of the platforms currently in extensive use. This line of work builds on the ideas that were already presented with regards to cross-national and regional, and cross-media scholarship. The vignette we included in the opening of chapter 4, on the emergence of Twitter Fleets and the reaction of the user community, shows the value of a historical gaze about the development of different platforms. In that brief history what emerged from the observation of design innovations in Twitter, WhatsApp, Facebook, Instagram, and Snapchat over time was a phenomenon of coevolution: The functionality of stories had emerged first in one platform to then be taken up in others in a process of mutual adaptation and ultimately convergence. To understand, for example, the ways in which Twitter users produced ironic memes about the functionality of Fleets, it was helpful to know that Snapchat had inaugurated the Snaps functionality years earlier and that in the meantime Facebook had added the stories functionality to its ecosystem of platforms. The historical perspective thus allowed us to make sense of cross-platform dynamics.

Another case that illuminates coevolutionary patterns is that of YouTube and Twitch. As we argued in the introduction to this chapter, playing online video games constitutes one of the most prevalent categories of practices on social media. Since its emergence in 2005, YouTube has been consolidating itself as one of the main platforms for observing gamers playing online. In its origins, the platform offered the possibility of sharing only recorded videos. Twitch was launched six years later, conceived as a livestreaming space associated with the media characteristics of television. In a short time, the video game genre became widely popular, and the platform began to attract millions of followers. In that same year, YouTube decided to launch its livestreaming functionality. For a long time, speculation swirled about YouTube's possible purchase of Twitch. Finally, the latter platform was acquired by Amazon in 2014.

It is common to find videos of YouTubers and streamers arguing why they decided to transfer from one platform to the other; in general, motivations have to do with the business model that each one offers to content producers. How do these digital diasporas shape the design and business strategies of YouTube and Twitch? To what extent has their relationship as competitors generated convergences and divergences in the technical possibilities they provide to users? These are questions that arise from comparison and that can be answered by taking a historical look at their development. The work we presented earlier by Kaye and colleagues (2021) proposed a view of Douyin and TikTok that arose from comparing business models of each platform in historical and geographical ways. Comparative work on platform coevolution aims to continue along this line on inquiry programmatically.

Conclusion

The scholarship on social media has tended to exhibit a strong present-day bias whereby research questions and objects of inquiry are situated within an "endless present tense" (Hartley 2018, 13). The examination of platforms that are widely used in current times is not inherently problematic; after all, understanding the social world includes making sense of contemporary phenomena. Furthermore, some of the most studied platforms have a tremendous reach nowadays; for instance, at the time of this writing, Facebook has 2.8 billion active users, or 37 percent of the world's population, which explains in part the contemporary focus. But focusing on the present without consideration of how we got to this stage has at least three limitations.

First, at the most basic level it makes it invisible how the past has shaped the present by both eliding the process of evolution of the platforms currently in use and neglecting the histories of platforms that were in use at some point and are no longer in existence. Second, the present-day bias entails artificially removing the current configurations and modes of use of platforms from broader cultural patterns that become easier to identify through historical accounts. Third, because both the platforms that are widely used today and the dominant ways in which they are used are those that have survived from a wide array of additional technical and practical options, overlooking the role of history implicitly moves success to the foreground and lack of success to the background. This, in turn, limits the analytical gaze to a relatively narrow set of objects of inquiry and runs the risk of misattributing the causes of their success; in other words, if we want to understand what makes a case successful, we also need to look at comparable cases that are not.

In this chapter we outlined a series of historical pathways to help overcome these limitations in cross-national and regional, cross-media, and cross-platform social media comparative scholarship. In the cross-national and regional dimension, we proposed the examination of how different national contexts shape divergent trajectories of the same platform over time, and the importance of de-westernizing social media genealogies. Furthermore, in the cross-media dimension, we suggested accounting for the influence of specific traditional media technologies and practices in the unfolding of different platforms and for the coevolution of traditional and social media over the past decade. Finally, in the cross-platform dimension, we highlighted the value of recovering the histories of dead or dying platforms and of continuing to explore coevolutionary dynamics—in this latter case, regarding how the main platforms currently in existence have mutually shaped each other.

Adding a historical sensibility to comparative work in social media scholarship helps to provide a more comprehensive account of how we got to where we are today. This entails shedding light on technologies, practices, and voices that have not made it to the mainstream, and unearthing forgotten—but not unimportant—milestones in the evolution of platforms today in the mainstream. As we suggested at the beginning of this chapter, platforms have multiple—and many times not self-evident—histories. Bringing them to the forefront of scholarly consciousness and integrating them into work that is contemporary focused contributes to making visible how the past might have shaped the present. This, in turn, enables the analyst to figure out areas of both continuity and discontinuity, thus showing what might be novel about platform dynamics—and what might be old wine in new bottles.

Undertaking comparative work with a historical mindset also brings to light the many ways in which contextual circumstances shape the trajectory and current state of platforms. Thus, the brief history of WhatsApp Pay's incorporation in Brazil acquires a potentially different meaning when learning about the history of the same platform functionality in India. Furthermore, recovering the contingencies behind the ascent of Silicon Valley in the world of digital technology contributes to both not taking its current configuration for granted and de-westernizing its role in global and local sociotechnical imaginaries. Moreover, contrasting the genre conventions and defining features of the dominant platforms with relevant conventions and features that were borne in the history of traditional media helps to contextualize the present.

Pursuing scholarship that incorporates a historical gaze within a comparative agenda also counters the limitations associated with the imbalance between platforms that are currently in use by hundreds of millions of people and those that either are no longer in existence or are used by much smaller portions of the population. Uncovering the many histories of dead or dying platforms not only enables the analyst to tell more comprehensive accounts of the past that are invisible through a present-day lens. It also provides important insights about the dynamics of lack of success in past times and, by implication, what might be the factors accounting for the success of the dominant platforms in the present. This, in turn, helps bring tension to the notion of novelty, particularly recurrent in discourses around social media technologies, and to embrace "the bleeding edge of obsolescence" (Chun 2011, 184) of all media and communication technologies. The novelty of media, as Ben Peters (2009) explains, implies a constant tension with the past; what is new is always transitional, so that their aspiration to newness is in a certain sense an impossible project. Incorporating a historical lens can operate as an antidote to the trap of obsolescence, thus avoiding a return to Groundhog Day in social media studies.

Finally, another implication from the project of de-westernizing the histories of social media ties to issues of language. Silicon Valley is not only a particular locale with a singular set of histories, but it is also a place where one language—English—dominates while others are relegated to the margins. To the same extent that Silicon Valley does not stand for all locations, English does not stand for all languages in which users engage with social media. It is to the pathway of language that we turn next.

6

Languages

Introduction

Aquí Se Habla Español

On October 30, 2020, Bad Bunny and Jhay Cortez release *DÁKITI*, a reggaeton song about the sexual tension of a relationship carried out in secret. Named after a beach in Puerto Rico, *DÁKITI* breaks audience records in record time. In less than a month after its release, it becomes the first song in history to reach the top of the Hot Latin Songs chart and the top ten of Billboard's Hot 100 at the same time. The video clip has more than one billion views on YouTube at the time of writing this chapter in April 2022.

On TikTok, the song is popular for dance challenges and viral lipsyncs. In one of the videos, with more than one million views, user @ralphlarenzo translates *DÁKITI*. The bio of the account states, "I sing Spanish songs in English/Yo canto canciones en español a inglés!" and this is accompanied by the Puerto Rican flag emoji. Many comments about the video revolve around the merits of listening to the song in Spanish versus in English. A user states, "Heck no, Spanish sounds better!!!" Another writes, "I'm glad its in Spanish lol." In response someone posts, "He's Puerto Rican—adding a skull emoji—it's not even Spanish it's Latin."

That popular translation of *DÁKITI* on TikTok and the ensuing dialogue is indicative of the tension between the global success of Benito Antonio Martínez Ocasio, a.k.a. Bad Bunny, and his decision to compose and sing all of his songs in Spanish—with the exception of his single *Yonaguni*, released in 2021, which has a coda in Japanese.[1] In his own words: "I write my songs, it's my ideas, my production, and I'm not going to have ideas and lyrics come to me in English. I've said it from the onset" (Cobo 2020). Whether intentional or not, this artist's position vis-à-vis English language can be read as an act of postcolonial resistance. Latinx and Latin American artists in the mainstream of the global music industry have been, at least until recently, expected to release versions of their songs in both Spanish and English. Bad Bunny notes that this linguistic practice was perhaps "necessary and they [the artists] opened doors to this Latin boom, but that moment for me is over. I am very proud to get to the level where we speak in Spanish, and not only in Spanish, but in the Spanish we speak in Puerto Rico. Without changing the accent" (Mars 2021; translation from Spanish).

The act of resistance and the vindication of identity, expressed with intensity in his reference to the issue of accent, is also evident in the artist's social media activity. Whereas many celebrities and influencers express themselves in English on their social media accounts to increase their appeal to a global audience, Bad Bunny—who has more than 38 million subscribers on YouTube, a similar amount of followers on Instagram, and 3.8 million followers on Twitter at the time of writing this chapter—writes all of his posts in Spanish. Yet, when it comes to how social media companies address this content, they sometimes translate it into English. In a video produced by the Twitter company and uploaded to its official YouTube channel, Martínez Ocasio is recorded saying, "Hey people, I'm Bad Bunny and these are the stories of some of my tweets" (translation from

Spanish).[2] Both Bad Bunny's words as well as the tweets he analyzes in the video are written in Spanish. However, Twitter adds English subtitles, thus positioning itself as an English-speaking platform. What language communities does a platform like Twitter imagine for its users to assume that a translation of a non-English video made by an artist who chooses to express himself in Spanish across his social media accounts is required and that such translation must be into English?

In a February 2020 interview with *Billboard Magazine*, conducted in English, Bad Bunny is asked about the name of his album *YHLQMDLG*. This album would end up becoming the first full-length Spanish-language album in history to top Billboard's all-genre chart. The dialogue that ensues is as follows:

Interviewer: Do you have a title [for the new album]?

Bad Bunny: Yeah, *Yo Hago Lo Que Me Da La Gana*.

Interviewer: Okay.

Bad Bunny: Okay? You don't even know what I said [laughing].

Interviewer: [laughing] I know, I cannot repeat that back, so that's why I had you say it for me, so that I didn't have to.[3]

In the context of this chapter, the dialogue is noteworthy for the power dynamics associated with the role of language, the tensions between English and Spanish, and the issue of translations. Who translates whom in a world in which English is often presented as the lingua franca of digital culture? When and why do these translations happen or not? How are they disputed? How does the assumption of English as the lingua franca of digital culture contrast with the multiplicity of linguistic experiences of the billions of social media users for whom English is not their native language or who do not understand this language? What does it mean, for instance, that Twitter exercises power in translating some, but not all, posts from some languages and into a few others? Paraphrasing Langdon Winner's (1980) seminal article, do social media translations have politics?

Like and Amen

In March 2013, Jorge Mario Bergoglio, born and raised in Argentina, is named the first non-European pope in history. After his appointment, Pope Francis takes over the Twitter presence of the papacy inaugurated by his predecessor, Benedict XVI. The number of followers rises dramatically since then—from three million in 2013 to more than fifty-two million at the time of this writing—to the point that Jorge Carrión calls Pope Francis "the first influencer appointed directly by God" (Solaris 2020). The millions of followers are distributed across papal Twitter accounts in nine languages, since Pope Francis tweets, almost simultaneously, in Arabic, English, French, German, Italian, Latin, Polish, Portuguese, and Spanish. The papal communication practices on Twitter end up reflecting part of the multilingualism inherent to the lived experiences of people using social media.

In 2016, Pope Francis's social media exposure expands to Instagram through the @franciscus account shortly after he meets with Kevin Systrom, then CEO of the platform; the text of each post is written in nine languages, in the form of a list. In November 2020, the account comes to the forefront of a public image scandal. Some followers notice that @franciscus has given, along with more than 132,000 other accounts, a "like" to a photo of Brazilian model Natalia Garibotto posing in a swimsuit. The event prompts the opening of an investigation inside the Vatican to trace back what many considered a serious error. However, beyond the content of the liked image there is another aspect especially salient for the current chapter: the signifier of the *like*. Besides the oddity of this particular like, since Pope Francis's official accounts never react to the content of other accounts, how is one to interpret the meaning of a nontextual signifier across multiple languages? Is it possible to imagine a universal interpretation of this type of signifier? Or, far from it, are we facing an iconography that is deeply polysemic and liable to generating misunderstanding within a millenary institution such as the Catholic Church?

In his message for the LIII World Communications Day, in 2019, Pope Francis ended up contrasting the significance of the like button in digital culture with the amen of the religious dogma: "This is the network we want. A net made not to trap, but to liberate, to guard a communion of free people. The Church itself is a net woven by Eucharistic communion, in which union is not founded on 'like' but on truth, on the 'amen' with which each one adheres to the body of Christ by welcoming others" (Pope Francis 2019; translation from Spanish).[4]

Just as the interpretation of sacred scriptures has led to a series of semiotic conflicts throughout the history of Catholicism, the appearance of new signifiers in digital culture has ushered social media practices into a terrain fraught with misunderstanding as the norm rather than the exception, making "the uncertainty of meaning" (Furedi 2016, 525) a topic of everyday conversation. When someone likes a post, is the post being liked at the level of the enunciation or at the level of the person who creates it? Is the like a sign of agreement, sympathy, irony, or of other intentions? Moreover, if the content being liked appears in different languages, is the meaning of the like also transformed by virtue of what liking might mean in different languages and within various national and regional contexts? Is the use and interpretation of this social media iconography universal, or is it a matter of global signifiers that are contextualized locally, mediated by each context's linguistic singularities and their cultural, institutional, and political associations?

Why Linguistic Comparisons Matter

One of the key constitutive elements of both subjectivity and social life, language is the second pathway we propose in this book to programmatically develop a comparative perspective in the study of social media. As with histories, its role is to make more robust the analysis of platforms in their cross-national and regional, cross-media, and cross-platform dimensions. This leads to denaturalizing an English-language bias that works, in many cases, in tandem with the present-day bias analyzed in the previous chapter. It also implies acknowledging the complex dynamics tied to novel visual signifiers that have become increasingly popular in social media in particular and in digital culture in general.

These two epistemic operations are aimed at countering two dominant scholarly practices that have marked accounts of language in social media within the field of communication studies. The first is the assumption of English as the lingua franca of digital culture (Mullaney 2017; Cheruiyot 2021; Mitchelstein and Boczkowski 2021; Suzina 2021). As Gerard Goggin and Jack McLelland argued more than a decade ago, "there has been little attempt to generate a discussion between scholars working on different language cultures or to develop modes of analysis that do not take Anglophone models as their starting point" (Goggin and McLelland 2009, 6). The second practice is the dominance of the textual dimension of platform use and the relatively much less attention given to visual signifiers whose polysemy resists the computational tools increasingly deployed to make sense of language as an aggregation of words and word frequencies (Highfield and Leaver 2016; Pearce et al. 2020).

The vignettes presented in the introduction to this chapter address these two topics and highlight some key challenges that emerge with the incorporation of a comparative linguistic focus into the study of platforms. The Bad Bunny vignette tackles the English-language bias and, in doing so, points to dynamics of oppression, resistance, and identity in the intertwinement of popular culture, social media, and politics. The Pope Francis vignette expresses, along with multilingualism, the complex place taken by novel visual signifiers—from likes to favs and from emoji to stickers. To counter the limitations posed by the English-language bias and by the dominance of text-only analyses, in this chapter we explore future paths of research in cross-national and regional, cross-media, and cross-platform scholarship that reveal the generative place of multilingual and multimodal communication in social media. Doing so will help us provide accounts with greater descriptive fit and heuristic power about platforms and their relationship with language.

Language in Cross-National Comparisons

Antecedents

Asaf Nissenbaum and Limor Shifman (2022) study the ways in which satire on social media in reaction to global events works across local cultures. Their goal is to "probe the multifaceted interactions between the global-local and entertaining-disruptive dimensions of contemporary digital satire" (Nissenbaum and Shifman 2022, 937). To do so, they compare a sample of humorous posts across Twitter and Weibo originally written in one of five languages—Arabic, Chinese, English, German, and Spanish—during November 2016 in relation to the election of Donald Trump as the forty-fifth president of the United States. Nissenbaum and Shifman (2022) find few themes and issues shared across satiric posts written in any one of the five languages under study. These were "references to physical appearance, personal relationships, and competitive political dynamics, none of which offered substantial criticisms" (937). Overall, global humor coincided in emphasizing entertainment, whereas locally oriented humor embraced the rather disruptive elements of satire.

Nissenbaum and Shifman also create a typology of local humorous responses to global events, distinguishing between "inbound," "transitional," and "outbound" satire across the geographic and cultural regions where the sample of humorous posts originated from. Inbound satire is characterized by analyzing global issues to compare them to local scenarios; this was central for the case of posts written in German. In transitional satire, which was frequent in posts written either in Arabic or Spanish, what matters is the symbolic position of the local audience and its relationship with global dynamics. In outbound satire, the local audience takes a detached position to comment about global events; this was noted to be mostly present in posts written in Chinese. In conclusion, their inquiry into language as used on social media across countries and regions allowed Nissenbaum and Shifman to illuminate dynamics of the interplay between globalization and political commentary.

Another study that demonstrates the potential of a language focus in comparative cross-national work is by Marco Toledo Bastos, Cornelius Puschmann, and Rodrigo Travitzki (2013), who center on transnational activism organized around specific causes. They ask whether language operates as a barrier or as a bridge when it comes to coordinating efforts through hashtags on social media. To do so, they examine a corpus of 8.4 million tweets and find that out of 455 hashtags, 53 percent were in English and the remaining 47 percent were predominantly in Portuguese or Spanish. The authors show that "linguistic division plays an important role in structuring the network communities" (Bastos, Puschmann, and Travitzki 2013, 166) and observe little overlap between groups that used different languages. However, they note that those hashtags linked to political activism, particularly around the *Indignados* and Occupy protests, had the highest level of network degree, suggesting that "political campaigns based on social media are driven by highly-active, politically engaged users that tweet across different hashtags and are immune to language barriers. Thus, political hashtags transcend linguistic communities, grouping together users and messages produced in a number of different languages" (Bastos, Puschmann, and Travitzki 2013, 168).

Drawing upon different theoretical and methodological frameworks, these two studies show how the incorporation of a linguistic perspective can highlight cross-national dynamics that are key to the lived experience of using social media: cross-cultural convergence and divergence (Nissenbaum and Shifman 2022), and barriers or bridges across linguistic communities (Bastos, Puschmann, and Travitzki 2013). However, despite their contributions, both studies lack an explicit cross-national agenda on language issues, which we begin to articulate next.

Future Developments

We propose two avenues for future research on language in cross-national social media studies. The first one concerns situations of territorial displacement, and the second one engages with the role of national and regional contexts in the production, circulation, and reception of new signifiers.

Approximately 3.5 percent of the world population reside in a country in which they were not born, exceeding predictions that were made almost two decades ago for the year 2050 (World Migration Report 2020). Furthermore, it is estimated that more than 1 percent of the global population is currently displaced either because of forced migration due to persecution or conflict, or because of statelessness.[5] For decades now, digital media have played a substantial role in articulating family and work communication across distances (Uy-Tioco 2007; Madianou and Miller 2013; Madianou 2016; Nedelcu and Wyss 2016; Gillespie, Osseiran, and Cheesman 2018; Madianou 2019), and in operating as instruments of surveillance and border control (Latonero and Kift 2018; Leurs and Smets 2018; Sánchez-Querubín and Rogers 2018). To address how this unfolds comparatively in social media, our proposed direction of research partly draws on work in the field of digital migration studies (Brinkerhoff 2009; Alonso and Oiarzabal 2010; Hegde 2016; Alencar 2018; Leurs and Prabhakar 2018). This field is dedicated to understanding the link between processes of territorial displacement and uses of information and communication technologies. Thus, it converges with prior scholarship that emphasized the central place of traditional media during diasporic experiences (Appadurai 1996; Karim

2003; Kraidy 2005), and also with the transnational turn of migration studies that focuses on studying ties across countries or regions (Faist 2004; Nedelcu 2012; Leurs 2019).

Situations of territorial displacement provide a unique window to explore dynamics related to the usual loss of the imagined or assumed correspondence between inhabited place and spoken language, thus challenging the notion that any language has a default status in social media practices, including English. That loss of correspondence is often tied to processes of power asymmetry that tend to be linked to already existing dynamics of economic and political inequalities. Therefore, these situations represent a significant terrain to examine the interconnections between multilingual practices and political dynamics as expressed on social media across national and regional settings. In addition, they highlight the issue of language diversity to which an individual who must move from one country to another is exposed.

Some possible topics to inquire within this future path of research would include how users with different national origins and in situations of geographic displacement manage switching languages when communicating through platforms with at least three different social groups: those remaining in their homeland, those who belong to their diasporic communities in other locations, and those colocated within the local territory. How do these dynamics play over time in cases in which groups of forced migrants are gathered in refugee camps that produce encounters of speakers of different languages? How do they differ in situations of voluntary migration? Furthermore, how do these dynamics vary in both cases in relation to the type of tie at stake, such as familial, romantic, friendship, and work, among others? What are the variations that relate to the kinds of topics discussed among speakers—namely, politics, religion, everyday life, sports, and work, among others? Finally, how are ties and topics influenced by the different power relations enacted in the various situations of territorial displacement across these three possible groups?

The second future avenue for research we propose centers on the role of national and regional contexts in the production, circulation, and reception of new signifiers that have become increasingly popular on social media. As discussed in chapters 1 and 2, a central tenet of common approaches to cross-national and regional studies has to do with assuming that national and regional contexts correspond to specific communication styles and practices (Hofstede 1983, 1991, 1998). Taking this into consideration, how have different contexts of origin shaped the ideation, design, production, and initial evolution of key constellations of new signifiers? For instance, what was the role of Japanese culture for the case of emoji (Gottlieb 2009), or South Korean culture for the development of stickers (Steinberg 2020), or culture in the United States for the design of reactions such as the "like" button (Bucher 2021)? To what extent have national variations in key occupational cultures—such as copyists, illustrators, graphic designers, engineers, and marketing specialists, among others—shaped the construction of these signifiers and the different options developed? In addition, how do these variations relate to broader visual cultural patterns such as the influence of manga in everyday life in Japan (Ito, Okabe, and Matsuda 2005) and the historical role of button technologies in communication interfaces in the United States (Plotnick 2018)?

One element that has made these new signifiers so powerful in digital culture is their cross-national circulation. That is, had they stayed popular only in their country of origin their influence in that country's everyday communication would have been great, but not so much at the global level—except perhaps within the practices of diasporic communities, as we noted earlier. Yet, these new signifiers have become a mainstay of social media partly because of their uptake in different parts of the world (Gómez-Cruz and Siles 2021). This opens a host of issues related to how they circulate in communication practices across national and regional borders. For example, what could we learn from the use and experience of WhatsApp groups if we considered the dynamics of sticker sharing among members situated in multiple countries with markedly different communication cultures? In addition, how do language choices in a smartphone's configuration shape which emoji are algorithmically suggested to users? Finally, does an apparently simple signifier such as the thumbs up emoji mean the same in national contexts with divergent cultures of interpretation? In other words, and going back to the second vignette we used at the start of this chapter, did the like by the @franciscus account mean the same to followers and parishioners in different parts of the world?

These last two questions point to the complex dynamics of reception in the case of the new visual signifiers, many times used to disambiguate the lack of tone that is associated with written language (Kavanagh 2016). Is a smile equally interpreted across national and regional contexts? The question exceeds the realm of social media, leading us to inquiring into whether the feeling of happiness and its gestural expression are universal or, in contrast, are decisively shaped by broader cultural configurations of everyday life different in various parts of the world. Cross-national and cross-cultural differences certainly apply when it comes to nonverbal communication (Lim 2002). Similarly, when the hashtag symbol is used to aggregate conversations on Twitter or Instagram, does belonging to these ad hoc communities of discourse mean the same to their participants located in different parts of the world? Finally, how are we to understand the various interpretations that users might have of emoji skin tones across places and their experiences and practices regarding issues of race and ethnicity?

In this section we began to develop the pathway of language by first drawing on studies that show the value of cross-national comparisons focused on linguistic variation, and second by outlining two concrete avenues for future programmatic scholarship: one focused on situations of territorial displacement and the other on the role of national and regional contexts in the production, circulation, and reception of new signifiers. However, language practices on social media often connect with those related to traditional media. It is to the examination of cross-media issues we turn to next.

Language in Cross-Media Comparisons

Antecedents

Rachelle Vessey's (2015) account of the coverage of the "Pastagate" shows the generative place of language in cross-media dynamics. This affair unfolded in Canada in 2013 when the Quebec Board of the French Language warned a local Italian restaurant to stop using terms such as "pasta" and "calamari" on its menu and to use their French equivalents instead to preserve this language in Quebec. Since the event produced a great deal of media coverage, both locally and internationally, Vessey analyzes how French and English languages were represented by traditional media in four countries—Canada, France, the United Kingdom, and the United States—and by user comments on Twitter. To do so, the author examines the language in which news stories and tweets were written, and their impact shaping the linguistic representations at stake.

Among other results, Vessey finds that both traditional English- and French-speaking media—except for Canadian French-speaking traditional media—and users' tweets largely produced negative representations of Pastagate, depicting "English as a humanised, international language that is necessary for business and French as a marginalised, overly policed language" (Vessey 2015, 268). The author notes that in a context of coexistence of traditional and social media, "the 'barometer' effect of the media reveals the intensification of pressure exerted on minoritised groups to translate linguistic cultures into English and globalised, market-driven contexts" (Vessey 2015, 269).

Another fruitful antecedent is that by Anna S. Smoliarova, Tamara M. Gromova, and Natalia A. Pavlushkina (2018). This study deals with an emotional aspect of news consumption—the use of Facebook reactions, a functionality that debuted in 2016 to accompany the "like" button, until then the only possible reaction offered per the platform's design. To do so, the authors focus on news consumption practices undertaken by the Russian immigrant community in Israel. They examine whether there is a correlation between type of reaction and the behaviors of either commenting or sharing a news story on Facebook.

Smoliarova and colleagues find significant correlations between the type of reaction and the kind of engagement with the news article. For example, posts with "angry" and "laughing" reactions tended to be more commented than shared, while those with the highest number of "likes" were not associated with any particular action. The authors warn that "the localization of verbal analogues may question comparative research of Facebook reaction usage across the world. For example, [the] Russian version of reactions that is studied in the paper includes 'outrageous' instead of 'angry,' 'super' instead of 'love,' and 'sympathize with/am sorry' instead of sad" (Smoliarova, Gromova, and Pavlushkina 2018, 251).

These two studies show the potential of looking at the role of language in cross-media comparative work. In Vessey (2015) we find tensions between English- and French-language representations that signal a degree of convergence across traditional and social media. The account by Smoliarova, Gromova, and Pavlushkina (2018) highlights how the use of new signifiers, which might vary across languages, tie to different types of interaction with traditional media content. Although both studies address to some extent issues of language across media, neither do so as part of a comparative programmatic agenda. We next continue developing it.

Future Developments

We outline two future paths of research in this subsection: first, we address the dynamics of translation across traditional and social media; second, we tackle the interactions across these two media regarding the incorporation of new signifiers in their respective language practices.

Traditional media with global reach, including leading news outlets and the film industry, have historically developed sophisticated translation processes to make their products available to consumers living in various parts of the world and communicating in different languages (Snell-Hornby 1999; Morley and Robins 2002; Straubhaar 2007; Lobato 2018). These processes have so far followed a relatively slow industrial and one-to-many logic, which has been recently disrupted by the relatively faster and many-to-many counterparts that are paramount on social media (Lacour et al. 2013; Lenihan 2014; Salameh, Mohammad, and Kiritchenko 2015; Desjardins 2016). Such disruption provides a fertile window to further our understanding of the role of language in cross-media dynamics.

Concerning production matters, it would be important to compare the human and technological resources devoted by traditional and social media to their translation efforts and the degree to which they are combined. While news and film companies have tended to rely mostly on human labor, social media companies have primarily resorted to algorithmic translation due to a combination of the volume of content available and the speed at which platforms operate. What happens when content originating in traditional media makes it to social media? How do translations happen in this process? How does technology shape it? Furthermore, what is the role of users-as-translators as opposed to that of professional translators hired by traditional media companies?

The last question leads into issues of distribution and circulation. For instance, although the design of platforms such as YouTube allows for a space to share song lyrics and user comments, the technology behind cable television channels such as MTV offers an environment, at least a priori, more resistant to multidirectional flows of information. Thus, when media products circulate in ways that cross boundaries between traditional and social media, they problematize these different stances regarding participation from the audience. The song *Yonaguni*, mentioned in the introduction of this chapter, surprised many Bad Bunny fans with its coda in Japanese. Within hours of its release, thousands of comments on the YouTube official video began to offer translations. As has become typical in the genre of "reaction videos," in which users share their takes on their first encounter with a media product, hundreds of people uploaded clips of themselves reacting to Bad Bunny's Japanese lyrics and sharing their interpretation of the coda. How do translations circulate in traditional versus social media? What are the various implications of such processes for the content's reach and reception?

Continuing with the reception of translations across media, how do consumers of traditional media interpret and engage with translated content versus the comparable processes undertaken by users of social media? Whereas the former have limited opportunities to voice concerns if they are unsatisfied with the translated content, the latter have ample avenues to not only express their dissatisfaction but also to propose alternative translations and make them available to other users. What are the implications of these divergent interactive capabilities regarding power dynamics between media production and consumption? Furthermore, the greater translation agency at the disposal of social media users has the potential to foster polysemy regarding the content that circulates, further illustrating the salience of the age-old trope of *traduttore traditore*[6] in the digital age.

The second path of future research we propose centers on how new signifiers so germane to platform communication have been incorporated into the language of traditional media and how the semiosis characteristic of traditional media has shaped language practices on social media. In which ways is the iconography embodied by elements like emoji, hashtags, stickers, and reactions represented in traditional media? Conversely, how do social media represent, in their digital environments, the visual and aural repertoire linked to traditional media, as can be seen, for instance, in disparate elements such as cinema billboards, radio jingles, television ads, and newspaper pagination?

In many cases, there seems to be a process of visual mimesis by which traditional media depict what they observe on platforms and further stabilize their meaning in popular culture. Stemming from the world of fiction, the series *Emily in Paris* (2020) tells the journey, from Chicago to Paris, of an American digital marketing expert primarily through the visualization on the television screen of the protagonist's Instagram account. In doing so, it presents an almost exact replica of the visual aspect of the platform, representing likes and reactions to signal the account's success in the character's life. In the domain of nonfiction, gossip and entertainment television programs and newsprint tabloids often draw upon the new signifiers as a source of scoops—for instance, by assuming from an exchange of likes the existence of a romantic relationship between two celebrities.

The comparative question about the representation of new signifiers in traditional media is particularly complex and far from settled. This is because of the strong polysemy associated with the repertoire of these signifiers on social media. The film *Searching* (2018) sets out to narrate a father's desperate search for the whereabouts of his teenage daughter entirely through screens in the digital environment. In its mimetic visual representation of platforms, it features scenes in which the meaning of a single emoji—for example, in Venmo, a social and mobile payment platform where emoji circulate to name banking transactions and socially smooth money exchanges—has the power to twist the course of a police investigation. How do traditional media manage this proliferation of new signifiers and their potential polysemy? How do they combine their own long-standing formats and languages with the recent but powerful emergence of social media iconography? How are the experiences of users on a given platform shaped by the presentation of what they see as emanating from traditional media?

In this section we continued developing the language perspective by building upon two studies that demonstrate the value of looking at language dynamics between traditional and social media, and then proposing two avenues for programmatic work: one focused on the dynamics of translation and the other on the incorporation of new signifiers. However, as we have argued repeatedly, social media are not uniform because there is significant heterogeneity across platforms. In the next section we delve into what this means for accounts of language practices.

Language in Cross-Platform Comparisons

Antecedents

A study by Michele Corazza and colleagues (2019) constitutes a useful instance of cross-platform comparisons focused on linguistic matters. The authors design and test a natural language processing methodology for detecting hate speech on social media able to operate in Italian across multiple platforms. Their motivation is that most of the existing data sets and approaches used to detect hate speech on social media are written in English and focus on one platform at a time, usually Twitter. To this end, Corazza and colleagues draw upon data in Italian from Facebook, Twitter, Instagram, and WhatsApp. They find, for example, that "learning to detect hate speech on the short length interactions that happen on Twitter does not benefit from using data from other platforms" (Corazza et al. 2019, 5). They also find that the emoji detection and transcription system is not as useful for this platform as for the others, probably because of the relatively lower use of these signifiers on Twitter. The authors conclude that "this shows that the language used on social platforms has peculiarities that might not be present in generic corpora, and that it is therefore advisable to use domain-specific resources" (Corazza et al. 2019, 5).

Noting that no previous studies have compared emoji use across platforms, Khyati Mahajan and Samira Shaikh (2019) examine this matter on Twitter and Gab, a platform heavily used by the alt-right community, especially in the United States. They analyze how emoji were used in content produced on these two platforms regarding the Charlottesville massacre, which occurred in the United States in 2017. The authors find that on Gab the sentiment was more negative, and that emoji were used in greater quantity than on Twitter. Furthermore, on Gab certain emoji linked to the political movement of former president Donald Trump (the frog face emoji) prevailed. In contrast, certain emoji that would indicate empathy, such as the broken heart or the peace sign, were more present on Twitter. Finally, Mahajan and Shaikh (2019) note Gab's use of positive emoji in a context of negative connotation, observing that "Gab users tend to use the emoji more in a sarcastic tone, whereas Twitter users tend to use the emoji more to express their disbelief during the event" (2).

Taken together, these two studies highlight the importance of undertaking cross-platform comparisons to understand language matters and show how far from settled content interpretations can be. The work of Corazza and colleagues (2019) showed that relying solely on English in a single platform would miss detection of hate speech in other languages and platforms, with important implications for the regulation of social media content. The study by Mahajan and Shaikh (2019) illustrated the variance in emoji use across two platforms. However, despite their significant contributions, none of these antecedents are part of a larger cross-platform comparative agenda. In the next subsection we continue the process of developing it.

Future Developments

We propose two possible research directions to advance a comparative agenda centered on exploring patterns of variation in written language and in the new signifiers across platforms.

The first direction inquires about the prevalence of various languages on different platforms. A common aspect throughout the design of platforms is that they tend to combine various degrees of personalization of the user experience, including the ability to configure preferred languages, together with different options of algorithmic translations into one or more additional languages. This opens the possibility of variation regarding the language or languages in which content is presented and also how it is received by users. The presence or absence of this variation, in turn, enables the analyst to probe a range of dynamics regarding culture, power, social structure, race and ethnicity, and gender, among others.

Does the country of origin of a platform affect the language considered official by the platform or the range of languages available to users to configure their settings? Furthermore, are there any recognizable patterns of variation by platform in this regard? Besides often having a default language and additional ones available in the user settings, some platforms offer the possibility of automatic translations of content posted in a particular language that an algorithm supposes the user does not understand. In this case, what are the criteria that influence algorithmic decision-making regarding translations of posts originally made in a language into another one, and how does this vary across platforms? Moreover, are these translations made visible (labeled as such), or do they remain opaque and therefore made invisible to users? Are these translations imposed or do they allow for a degree of customization by the user? How does the translation rating system shape the service offered? Again, how do these variations across platforms affect the dynamics of content production, circulation, and reception? What happens when the same company owns a constellation of platforms, such as the case of Meta's ownership of Facebook, Instagram, and WhatsApp? How do linguistic and translation policies vary across them? In the Tower of Babel of social media (Mocanu et al. 2013), questions such as these can help illuminate variations in language production and distribution across platforms that can in turn help analysts address broader cultural, social, and political matters.

In addition to these issues of language variation in production and distribution dynamics, future research could also inquire into language use and interpretation. Are there major patterns of variation in terms of the languages used across different platforms? Furthermore, do different language communities form within platforms? It is not uncommon, for example, to find YouTube or TikTok comments from users who wish to gather around their own imagined linguistic community, through messages such as "where are the ones who speak [insert language]?" How does this vary, if at all, by social media platform, and why? Beyond the language or languages commonly used by platform, there are issues of interpretation. How do users who speak different languages make sense of a post originally made in a language different from theirs, and how does this vary by platform? Do they share their interpretations publicly on platforms? Moreover, what do they think of the aforementioned algorithmic translations—specifically, those that platforms offer them often in the absence of users requesting them—and also regarding their resulting quality?

The second future path of research proposes inquiries into the cross-platform variation surrounding the repertoire of new signifiers prevalent on platforms. This iconography is expressed linguistically on at least two levels: textual and visual. Textually, a series of questions arises regarding the ways in which the new signifiers are translated into different languages and the implications that such translations have on the production, circulation, and reception of the content. For example, when it comes to reactions, how are their official names translated into different languages on various platforms? As the work of Smoliarova, Gromova, and Pavlushkina (2018) already suggested, these differences lead to important questions: To what extent is it possible to compare the use of Facebook reactions across languages in which these signifiers have different linguistic value? For example, in Spanish there are at least two expressions to communicate love for someone or something—*te quiero* and *te amo*; in English there is only one that dominates—I love you. The possible range of Facebook reactions in Spanish does not seem to contemplate this difference, opting for a third option that is placed as the equivalent of love in English and that switches from noun (object) to verb (action): I love it [*me encanta*].

Visually, social media have been the seedbed of a new iconography of signifiers, from vernacular signs original to a particular platform, such as the at sign or the hashtag on Twitter, to elements shared across multiple platforms, such as emoji or stickers. In terms of their production, do these repertoires vary by platform? How do representations of the same signifiers change visually depending on the platform and its interaction with the operating system of the device in which users access the content? When it comes to their circulation, in what ways do certain signifiers travel from one platform to another? How are their uses stabilized or contested, in a context of potential polysemy? On Twitter the hashtag often groups a content within a series of discussions, but on Instagram the same symbol is generally used to increase the visibility of a post. Finally, in imagining their reception, how does the use of new signifiers vary across platforms? Why are certain social media, as Mahajan and Shaikh's (2019) work demonstrated, more prone to frequent emoji use than others? In what ways does the sharing of new signifiers across platforms transform over time?

The exploration of language patterns across platforms leads to a number of questions and avenues for research since it enables the analyst to explain and understand aspects of social media use that remain, in many cases, invisible or relatively little discussed. In this chapter we have suggested several possible directions of research in cross-national and regional, cross-media, and cross-platform dimensions aimed at building a programmatic agenda of future work. Next, we close this chapter by bringing these various strands together.

Conclusions

We have argued that just as the pathway of histories aimed to counter the present-day bias that runs through much of the literature on platforms, the pathway of language intends to offer an alternative that complements the twin tendencies to take English and textual communication as the default modes of symbolic praxis in scholarship on social media. The vignettes presented in the introduction sought to illustrate these tendencies. Bad Bunny and his use of social media in Spanish, contrary to the translations imposed by different platforms and the global music industry, underscored the intersection among language, politics, and popular culture. Pope Francis and his platform practices not only showed a strong multilingualism but also highlighted the polysemic conflict provoked by a novel signifier such as the like button. Moreover, throughout the middle sections of this chapter we sought to broaden the spectrum of languages and visual signifiers that would be helpful to study on platforms.

We proposed three avenues of research to counter the English-language bias—one for each of the three dimensions of comparative work we address in this book. Regarding cross-national and regional studies, we suggested examining the use of social media in contexts of territorial displacement, both voluntary and forced, in which a high degree of linguistic diversity tends to be present. As Sirpa Leppänen and Ari Häkkinen (2013) have argued, "Within them [social media], communication and interaction are often multimodal and linguistically and discursively heterogeneous, such heterogeneity serving participants as a means for indexing identifications which are not organized on the basis of local, ethnic, national or regional categories only, but which are increasingly translocal. In social media practices, participants are thus orienting not only to their local affiliations but also to groups and cultures which can be distant but with which they share interests, causes or projects" (2013, 18).

Moreover, for cross-media scholarship we outlined a research direction focused on the processes whereby translations are produced, circulated, and received in interactions between traditional and social media. Finally, in terms of cross-platform studies we argued that accounts of variance of languages used in the different platforms could constitute a particularly fruitful terrain to explore larger societal issues at play in both continuities and discontinuities of experience across the ever-growing array of platforms that constitute the social media landscape.

As Barton and Lee (2013) argue, "instead of examining CMC [computer-mediated communication] from a solely monolingual, usually English, perspective, a growing body of research is interested in how speakers of various languages have adopted such new forms of writing to different extents" (6). Thus, we also suggested three avenues of research to complement the dominant textual focus of most social media scholarship and built on contributions from the domain of digital discourse studies (Herring 1996; Thurlow and Mroczek 2011; Thurlow 2018; Bou-Franch and Garcés-Conejos Blitvich 2019; Sumner et al. 2020). Regarding the cross-national and regional dimension, we outlined a series of strategies to examine the production, circulation, and reception of novel signifiers in different geographic locales. In addition, to further develop cross-media comparative work, we offered alternatives to inquire into how the novel signifiers are represented and used in the context of traditional media, and how social media represent and use visual elements specific to traditional media. Finally, concerning cross-platform studies, and in consonance with our approach to the dominance of English in accounts of textual communication, we foregrounded the use of novel signifiers in different social media.

Language is central to the constitution, in a relational fashion, of both personal and collective experiences: as Ferdinand de Saussure ([1916] 1983) argued in his seminars taught over a century ago, the value of specific words emerges from interactions among signs that are contextually dependent. Applied to the study of social media this means that signifiers acquire different meanings in relation to other signifiers in the eyes of users situated in various parts of the world, encountering them on multiple media, and on diverse platforms. Because we signify comparing and we compare signifying, the study of language is an ideal way to build the epistemic perspective we advocate in this book.

7

Blurred Genres, Trading Zones, and Heterogeneous Inquiries

The journey of this book began with an organizing epistemological principle: to know is to compare. Based on the realization that studies of social media have generally lacked comparative perspectives in at least three dimensions—across nations and regions, across media, and across platforms—we set out to develop a research agenda exploring productive intersections across the foundations of different scholarly traditions to make sense of how platforms are designed, developed, circulated, and used. To this end, we drew inspiration from Clifford Geertz's notion of "blurred genres" (1980) to explore common logics of comparison across these different dimensions and their associated scholarly traditions.

The next section takes stock of what we learned from this intellectual journey about each dimension of comparison and our deep dive into two pathways for future development—namely, histories and languages. Our account reveals a thread weaving seemingly disparate strands of scholarship: the heterogeneity of social media. This heterogeneity was already conveyed by our selection of the vignettes with which we opened chapters 2 through 6. Taken together, these vignettes combined spheres of social life as disparate as cross-national diffusion of public protests against gender and racial oppression;

cross-media relationships in reality television and political journalism; cross-platform patterns of imitation and regulation; historical dynamics behind moral panics and multiple genealogies; and the resistance against English-as-default and the potential of multilingualism on platforms. We argue that this heterogeneity characterizes both social media as objects of inquiry and the ways in which they have been studied—the latter issue to be addressed more fully in the other two sections. This, in turn, presents important implications for the comparative turn we advocate in this book.

In the following section we probe the issue of heterogeneity further by exploring the potential integration of scholarship across two or more dimensions. We account for why we have discussed each dimension separately in the previous chapters and explore what it would entail to pursue work that integrated multiple dimensions. At stake, we argue, is the construction of accounts countering the intellectual fragmentation that has characterized the study of media and communication more generally—a fragmentation that some see as increasing in recent years. This state of affairs signals both the possibilities but also the challenges of blurring genres of scholarship.

In the final section of this chapter, we reflect on what a focus on comparative perspectives can contribute to addressing this fragmentation. We contend that the deep historical roots and contemporary intensification of this trend indicate a fundamental disunity in the relevant scholarship. Furthermore, we show that neither the existent theoretical nor methodological attempts to bridge different subfields and their associated traditions of inquiry have succeeded at reducing this trend. Thus, we propose that the epistemological turn to comparative perspectives can create trading zones (Galison 1997) across the various traditions of inquiry—about social media matters in particular, and other media and communication phenomena more generally—where local coordination about specific research pursuits can take place without flattening the larger heterogeneity of the field.

The Heterogeneity of Social Media

Our articulation of the organizing principle that animates this book—to know is to compare—has yielded a series of insights that helped us understand the conditions of existence of social media in global, transmedia, and multiplatform communication environments. None of these insights would have emerged without a comparative approach since it was its application what allowed them to become more visible and put them within the context of larger scholarly conversations. The studies examined throughout the book shed light on topics generally addressed by social media scholarship—such as identity making and self-presentation, relationship maintenance and social capital, and political participation and activism. They were also able to illuminate variations across structural dimensions of the production, distribution, and use of platforms, such as the dynamics of racial and ethnic discrimination, the platform economy of social media, and the logics of datafication and algorithmic bias. In the following paragraphs we will summarize these insights and weave them into an argument that emphasizes heterogeneity as a key feature of both social media and the scholarship about them.

The first insight that emerged from applying the comparative lens to cross-national and regional phenomena is the importance of examining the relationship between the nation-state, globalization, and social media. This relationship is contingent on how social media operate in different national settings. As we argued in chapters 1 and 2, traditional media have played a key role in the constitution of nation-states. The latter, in turn, have shaped the former in various ways. Perhaps paradoxically, the technological evolution of traditional media, together with the emergence of so-called new media, have also played a key role in processes of globalization. In turn, globalization has called into question the importance of the nation-state: If the existence of almost 200 countries in the world is a testimony to political, economic, and cultural heterogeneity (among other dimensions), globalization embodies the countervailing tendency toward greater international homogenization.

We argue that to understand how social media are defined, constructed, circulated, and used in different countries, it is essential to unravel on a case-by-case basis the place occupied by the nation-state. Platforms are objects with a potential global reach where the role of the nation-state seems in many cases to be blurred—think, for example, of issues regarding the regulation of hate speech and violent content addressed in chapter 4. Thus, theoretical debates have arisen, as we explored in chapter 2, about whether the country of origin or use should be considered a significant explanatory factor in the first place, with potentially important implications for understanding heterogeneous dynamics of racial and ethnic discrimination within platforms, among other topics. Yet, throughout chapter 2 we showed how in some situations there was substantial divergence in social media use that could be attributed to the persistence of variables associated with the nation-state, whereas in other instances there were phenomena common across borders. Following Livingstone (2003), we concluded that the nation-state appears in principle to remain an important factor, but one whose presence nonetheless cannot be assumed. Thus, we proposed that the role of the nation-state in each case should not be taken for granted but demonstrated as an outcome of the research process.

The second insight was the relevance of traditional media for understanding the genealogy of their digital counterparts, including platforms. Comparing social media with preexisting or coexisting traditional media reminds us that, as Lisa Gitelman (2006) has argued, media exist in a plurality. This plurality reinforces the idea that the media ecosystem has long been heterogeneous. But such heterogeneity does not necessarily imply either the absence of bridges between the new and the old or the assumption that novelty is always associated with a break from what came before. On the contrary, the application of the comparative approach allowed us to note both a series of continuities between traditional and social media as well as discontinuities between them, which has been key to contextualizing, for instance, the relationship between the platform economy of social media and the commercial logic of traditional media.

Moreover, as we pointed out in chapters 1 and 3, traditional media have not only contributed to shape new media, but the latter have also remediated the former, with the resulting pattern that all media coexist in ecosystems marked by their heterogeneity. Like the case of the nation-state, this means that just as the discontinuity between social media and their traditional media counterparts cannot be assumed, neither can it be assumed that there is always historical continuity. Thus, heterogeneity in cross-media dynamics invites us to interrogate them and demonstrate whether continuity or discontinuity applies in each case.

The third insight is that social media are plural also because they include a wide range of platforms with distinct genealogies, technological properties, and cultures of use. This stance challenges the dominant mode of social media research that has tended to focus on one platform at a time—often privileging options such as Facebook or Twitter. This dominant mode has also usually been enacted without reflecting on how its view from nowhere assumes a certain homogeneity across platforms that does not reflect their various logics of datafication as well as their actual use. If all platforms were similar, it would be hard to understand why, as mentioned in chapter 4, users on average were active on more than seven platforms in 2022.

On the contrary, the cross-platform comparative perspective enabled us to delve into the notion of social media as an object of inquiry that is heterogeneous by definition and therefore far from constituting a unified whole. Throughout a series of studies, we noticed, for example, how users produced complex differentiations between platforms that a priori presented similar technological capabilities. This, in turn, led us to question dystopian discourses that focus almost exclusively on the algorithmic and commercial design of each platform and its presumed impact on society. Looking at the heterogeneity of platforms increases the conditions of possibility for identifying cross-platform variation. This invites us to explain both the presence and absence of dissimilar effects and of negative and positive social consequences. By implication, this stance envisions dystopian discourses not as an a priori of research but as a symbolic formation that should emerge—or not—as a result of it.

The fourth insight about issues of heterogeneity stems from adding a historical sensibility to comparative perspectives. This enabled us to counter the predominant present-day bias that has produced at least three effects: erasing the evolution of social media over time; blurring the ways in which many of its current features find antecedents in past media; and reifying success stories while neglecting the lessons that arise from recovering the histories of platforms no longer in use. Weaving through the cross-national, cross-media, and cross-platform dimensions, the historical gaze expands the lens of which phenomena are relevant to describe and which factors appear to explain their trajectory over time.

This gaze also reinforces the centrality of heterogeneity in the study of social media. Platforms can evolve in relation to the national contexts in which they are built and used, to the traditional media that precede them and with which they coexist, and to the other platforms with which they compete for users' attention. Furthermore, they can also change in their temporal evolution. Facebook today is not what it was a few years ago, nor what it will be in a few years—assuming it continues to exist. This historical sensibility leads us to explore continuities with the past that help us better assess any discontinuous aspects in the present. But above all it invites us to delve into the various combinations between determination and contingency that characterize the passage from what was to what is. It also reminds us that what we think will become in the future from the point of view of the present is not something inevitable but a conjecture among other possible ones.

The fifth insight resulted from our interrogation of a central element of symbolic and relational life: language. Proposing a comparative approach that took language as its focus opened the possibility of countering two dominant biases in studies of social media. First, the English-language bias, whereby the production, circulation, and consumption of platforms are by default imagined as configured and experienced in the English language. Second, the written-text bias, whereby social media communication should be understood in a written-textual key, leaving aside the emergence, stabilization, and centrality of visual languages and new signifiers that increasingly characterize symbolic expression on platforms.

The journey through cross-national, cross-media, and cross-platform comparative perspectives that centered on the role of languages enabled us to make visible the heterogeneity of linguistic realities linked to the lived experiences of users and to movements within and across nations, media, and platforms. This heterogeneity of languages and signifiers contrasts sharply with the homogenous view centered on the English language and on textual communication as the dominant modes of symbolic expression on social media. This, in turn, transforms issues of choice of language and mode of signification into questions rather than taken-for-granted answers. This is not to say that in some (or many) cases, social media communication does not occur in written English and in textual form, but that even in these cases the existence of such communication is a phenomenon to be explained rather than a premise to be accepted without interrogation.

In sum, foregrounding the heterogeneity of social media and their scholarship through adopting comparative perspectives, and probing this heterogeneity with particular intensity regarding matters of histories and languages, invites us to turn assumptions into questions and certainties into conjectures. A further step in this direction occurs when we aim to integrate multiple comparative perspectives, an issue we discuss next.

Integrating Multiple Comparisons

Throughout the book we have opted to discuss comparisons across nations and regions, media, and platforms separately. This decision was informed by argumentative, intellectual, and institutional factors that align with the heterogeneity of both social media as objects of inquiry and the ways in which their study has often proceeded.

From an argumentative standpoint treating each type of comparison in a distinct way has helped us articulate what it consists of, how it differs from dominant modes of scholarship, which contributions it enables the analyst to make, and what broader theoretical issues it illuminates.

From an intellectual perspective each of these modes of comparison has been partly shaped by varying traditions of inquiry which have historically evolved into semi-autonomous communities of discourse. Thus, cross-national and regional comparisons build on prior institutional and historical accounts in political communication and journalism, intercultural approaches to communication, and cultural analyses of global media patterns. Cross-media comparisons are informed by insights from institutional and cultural perspectives on media evolution and from historical analyses of technological change in information, communication, and media artifacts. Finally, cross-platform comparisons have drawn from a combination of audience research and cultural approaches to technology use.

In addition to the role played by these different traditions of inquiry, social media comparative work along these different dimensions has focused on divergent topics, examined them with different approaches and methodologies, and interpreted the resulting findings from varying lenses. Thus, while cross-national and regional accounts of social media have often concentrated on topics such as ideological polarization and political debate, cross-media work has recurrently examined issues such as the relationship between media and politics or journalism, and cross-platform scholarship has frequently delved into matters such as interpersonal communication and presentation of the self and the impact of platforms on mental health. Furthermore, whereas work in cross-national and regional studies has typically been marked by political economy or intercultural approaches, both cross-media and cross-platform comparisons have often featured perspectives emphasizing continuous and discontinuous patterns happening either synchronically or longitudinally, thus featuring various evolutionary and coevolutionary dynamics.

Methodologically, cross-national and regional studies have often conducted large-N studies based on surveys and—albeit to a lesser extent—carried out small-N data analyses. In contrast, cross-media and cross-platform studies have resorted to both quantitative and qualitative techniques such as surveys, experiments, network analyses, interviews, focus groups, and discourse analyses. Finally, in cross-national and regional accounts the interpretive lenses have often revolved around issues of convergence and divergence among media systems, political institutions, and cultural configurations. In contrast, in cross-media studies these lenses have frequently been organized around dynamics of reinforcement and displacement between social and traditional media. Moreover, in cross-platform work scholars have typically resorted to lenses that prioritize explaining either why different affordances of specific platforms produce divergent effects or why modes of appropriation vary beyond similarities in technological design.

In addition to these argumentative and intellectual factors relevant in addressing issues of heterogeneity, the decision to discuss the three dimensions separately has also stemmed from institutional matters. The multiple traditions of inquiry associated with the different types of comparative scholarship about social media map onto various subfields in the study of media and communication. These subfields, in turn, are linked to different divisions, sections, interest groups, preconferences, and workshops in professional societies; journals and book series in the publishing space; and jobs and curricular developments within colleges and universities. As the field has become increasingly specialized and expectations of publication volume have risen in recent years, the intellectual distances across these multiple institutional expressions appear to have widened further, thus leading to a state that Silvio Waisbord (2019) has recently characterized as "intellectual fragmentation." It has become common that scholars working on cross-national matters rarely engage with the concerns of studies on cross-media and cross-platform dynamics, and vice versa. Although the increased specialization has partly been responsible for a significant growth in scholarly output, the ensuing fragmentation has sometimes artificially severed connections that could enable analysts to present a more nuanced and holistic understanding of their object of study.

This fragmentation is both an expression and a limitation of the heterogeneity of social media and the resulting value of developing perspectives that can account for it. Thus, in addition to stressing that dynamics taking place in one country are not necessarily similar to dynamics present in other national settings, that what applies to one medium does not apply by default to another medium, and that phenomena specific to one platform in many cases are not replicated on other platforms, throughout chapters 2 through 6 we have highlighted connections across the different dimensions of comparative work. Building on this, in the reminder of this subsection we explore the potential of integrating comparative analyses across two or more of the dimensions analyzed in previous chapters. More specifically, we visualize this integration on Figure 7.1 through a Venn diagram that maps the distinct intersections among

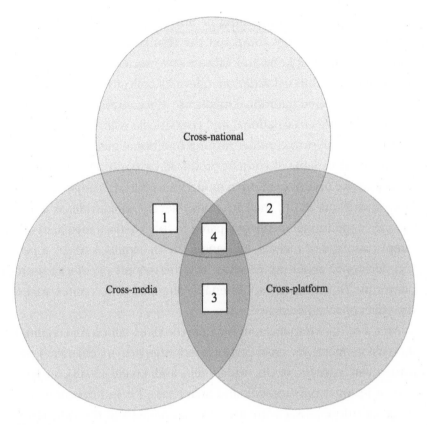

Figure 7.1
Distinct intersections among the cross-national and regional, cross-media, and cross-platform dimensions.

the cross-national and regional, cross-media, and cross-platform dimensions.

Area 1 of the Venn diagram is where the cross-national and regional dimension meets its cross-media counterpart. In chapter 2 we presented a paper by Skoric, Zhu, and Lin (2018) on political unfriending or unfollowing on social media. The authors asked, among other things, whether social factors such as the degree of collectivism in a society or psychological factors such as the experience of fear of missing out (FOMO) were somehow linked to the practice of political unfriending or unfollowing on Facebook and Twitter. To do so, they compared the results of an online survey conducted in Hong Kong and Taiwan—two societies that according to the authors exhibited variation in level of collectivism and FOMO. The authors treated traditional media use as a control variable. However, integrating cross-national and cross-media comparisons would inspire the analyst to consider turning traditional media use into an independent variable. If content produced by traditional media is a central source of conversation in modern societies (Harrington, Highfield, and Bruns 2013), then it seems relevant to consider whether the use of traditional media compared to that of social media, in different countries or regions, influences the dynamics whereby a person decides to avoid the presence of other people on one or more platforms. This, in turn, would enable the analyst to contextualize the effect of social media per se.

Area 2 of the Venn diagram is where the cross-national or regional dimension meets its cross-platform counterpart. In chapter 4 we introduced a study by Utz, Muscanell, and Khalid (2015) on feelings of jealousy on Facebook and Snapchat. The authors compared the association between the use of social media and the experience of jealousy in the context of romantic ties. Analyzing data from an online survey of participants mostly from England and Scotland and, to a lesser extent, from countries not identified by the authors from Europe and beyond, Utz and colleagues (2015) found that while platform use did not seem to generate high levels of jealousy, the effect was greater on Snapchat than on Facebook. They attributed this difference to the fact that on Snapchat communication appeared to be more intimate than on Facebook. Despite having collected data from different countries, however, the authors failed to explore the possibility of cross-national variation in the findings. Since social norms associated with romantic relationships and emotional states have been shown to vary by national setting (Bhugra 1993), it would have been worth exploring the relationships between these aspects of platform use by country. Both the existence and absence of cross-national variations in platform use would have been interesting findings whose explanations could have contributed to theorizing the intersection between cross-platform and cross-national dynamics.

Area 3 of the Venn diagram is where the cross-media dimension meets its cross-platform counterpart. In chapter 3 we discussed a paper by Rebecca Nee and Valerie Barker (2020) on the social impact of coviewing in second screening situations. The authors inquired into the capacity of second screening to provide a sense of togetherness when carried out both in relation to traditional and social media. Nee and Barker (2020) found more similarities than differences in coviewing experiences with respect to traditional media content and YouTube. Integrating cross-media and cross-platform foci would invite the analyst to also include second screening experiences with respect to content viewed across different social media platforms. Contrary to the more common way of framing second screening as the experience of consuming either traditional or streaming media and commenting about it on social media, an integration of dimensions would lead us to inquire about the conversation that happens across different platforms about the content that is encountered primarily on social media. This, in turn, would allow us to compare the social effects of interaction about content in traditional media versus on different platforms.

Area 4 of the Venn diagram is where all three comparative dimensions converge. In chapter 4 we introduced a study by Ariadna Matamoros-Fernández (2017) on the circulation of racist speech in Australia on Twitter, Facebook, and YouTube. The author found, among other things, that the posts tended to escape moderation controls. Moreover, recommendation algorithms usually reinforced the circulation of this type of content. While Matamoros-Fernández's work focuses on the cross-platform dimension, an integrative view of cross-national, cross-media, and cross-platform dimensions would, for example, allow us to understand the transnational reach of platformed racism as well as to compare the forms of moderation of racist content in traditional versus social media.

We have used this Venn diagram to map different ways of integrating the three comparative dimensions that we have mostly explored separately in previous chapters. Integrating these comparative perspectives has enabled us to make visible potentially relevant aspects of inquiry that would have remained less visible otherwise. Doing so invites us to pose new questions, identify significant sources of variability, contextualize observed phenomena, and increase explanatory and interpretive power. Yet, despite this potential, the paucity of attempts to undertake this kind of scholarship is built on a long-standing divergence among traditions of inquiry, as noted earlier. Thus, in the next section we close this chapter by arguing that the comparative perspectives proposed in this book have the potential to create epistemological trading zones helping to bridge the heterogeneity of studies about social media and about other media more generally.

Comparative Epistemologies and Trading Zones

The study of media and communication has been historically characterized by its heterogeneity. In the Anglo-Saxon arena there have been repeated accounts of this heterogeneity over the years.[1] One manifestation of this trend has been the proliferation of special issues devoted to matters such as "fragmentation," "ferment," "intersections," and "speaking across subfields" published by *Journal of Communication*, the flagship outlet of the International Communication Association, in 1983, 1993, 2008, 2018, and 2020, among many other related special issues and edited volumes (Gerbner and Siefert 1983; Levy and Gurevitch 1993; Pfau 2008; Fuchs and Qiu 2018; Tenenboim-Weinblatt and Lee 2020). Wiemann, Pingree, and Hawkins (1988) have gone as far as saying that the study of media and communication had been "bisected before it was united" (307). Yet, despite critiques about the potential deleterious consequences of these splintering dynamics, scholars have recurrently noted that these dynamics appear to have deepened. Already three decades ago Charles Berger (1991) asserted that "the traditionally high level of fragmentation manifested by the field seems to be increasing as the field expands" (101), a claim that (as noted above) Silvio Waisbord (2019) echoed in his account of these matters partly inspired by his experience as editor-in-chief of *Journal of Communication* from 2010 to 2016.

Scholars have often underscored that this heterogeneity is linked to an intellectually parochial predisposition. Michael Pfau (2008) observed that "the tendency is for scholars to burrow deeper into their respective niche, treating their own specialty as if it were isolated and self-contained" (599). This kind of intellectual parochialism resonates with the divisions among patterns in scholarship on social media that we identified in chapter 1. Thus, repeated calls for approaches aimed at overcoming this tendency are unsurprising. One of the most recent calls was made by Keren Tenenboim-Weinblatt and Chul-joo Lee (2020) in their introduction to a special issue about scholarship across subfields: "[C]ross-cutting discussions and integrations are crucial for theoretical innovation, for fuller and deeper understanding of communication processes and effects, and for the field's ability to achieve public impact" (304). Thus, scholars have most commonly attempted to engage in these discussions and integrations through either methodological or theoretical strategies.

The focus on methodological strategies in part stems from the centrality of attention to matters of method among many students of media and communication phenomena—which, according to Charles Berger (1991), sometimes has verged on an "almost obsessive preoccupation" (105). Within scholarship on social media, computational techniques have been the main candidate for integrative approaches in recent years. This builds on the high expectations that many researchers across the behavioral and social sciences—and to a lesser extent the humanities—have placed on this kind of techniques (Lazer et al. 2009; boyd and Crawford 2012; van Atteveldt and Peng 2018; Wagner et al. 2021). Along these lines, commenting on the submissions received for their recent special issue Tenenboim-Weinblatt and Lee (2020) state that "in communication research, the most notable development over the past decade has been the rise of computational methods" (305).

Despite these high hopes and the equally high level of popularity in some quarters, we argue that computational methods have significant limitations to foster productive conversations among disparate traditions of inquiry in social media scholarship—and other media and communication phenomena—because of their inherent difficulties in capturing the manifold global, transmedia, cross-platform, historical, and language dynamics examined in chapters 2 through 6. That is, despite the advances in computational power that undergird the contemporary renewal of expectations to develop a universal language of observation already articulated by the philosophers of the Vienna Circle a century ago, the complex differences in meaning and practice that emerge in relation to multiple national, media, and platform settings; the often circuitous unfolding of these matters over time; and the challenges in dynamics of translation and signification that resist the transition from words and images to numbers, posit limitations to the ability of computational methods to foster productive conversations across the various dimensions of scholarship on social media. In addition, the deployment of computational methods as a way of bridging subfields would require methodological homogenization across disparate traditions of inquiry, something that would most likely elicit significant levels of intellectual and institutional resistance among scholars and units with different methodological inclinations.

Deploying theoretical resources to foster these cross-cutting conversations has also proven to be quite limited, albeit for two reasons

that differ from the use of methodological strategies. The first reason is because the most popular theories in the study of media and
communication have historically been tied to particular subfields
and have been applied much less frequently outside of them. Even
those developed to account for social media phenomena tend to
remain circumscribed to their domain of application. For instance,
even though the definition of "context collapse" (Marwick and boyd
2011) builds on Joshua Meyrowitz's work on traditional media reception (1985), the majority of the work engaging with the concept with
respect to social media does not engage in cross-media accounts,
which is consistent with our argument in chapter 3.

The second reason is the remarkable stability of the most popular
theories used in the study of media and communication. As Walter,
Cody, and Ball-Rokeach (2018) conclude in their study of scholarship
published in *Journal of Communication* over the past six decades, a
handful of theoretical frameworks has largely dominated the explanations offered by scholars: framing, agenda setting, social learning
theory, narrative theory, uses and gratifications, and so on. As we
have shown in chapters 2 through 6, these are some of the main theoretical resources also used to account for social media phenomena
from comparative perspectives. Therefore, if the use of these theoretical frameworks by itself has not been enough to counter fragmentation until now, there is no reason to expect it would have
that effect in the future.

In contrast, we propose that the adoption of comparative perspectives as a key epistemological principle organizing scholarship
on social media is a more fruitful alternative than primarily methodological or theoretical strategies to engage in cross-cutting work. This
is because these perspectives have the potential to become something
akin to the "trading zones" identified by Peter Galison (1997) in his
account of the productive exchanges between theorists, experimen-
talists, and instrument makers in modern physics. There, Galison
shows how scientific communities that differed in a number of key

theoretical, methodological, and institutional matters were nonetheless able to fruitfully exchange critical key resources in a highly localized fashion that did not require coming to consensus about larger intellectual matters. Thus, "the focus is on finite traditions within their own dynamics that are linked not by homogenization, but by *local coordination*" (Galison 1999, 145; emphasis in the original).

This local coordination takes place in a trading zone where the different parties meet because, as Pamela Long (2015) has suggested, "each party has a particular knowledge or skill that the other side values as something they would like to possess or use in their own work or thinking" (843). The exchange of knowledge does not require either significant intellectual compromises among the parties or broad translations of methodological and theoretical ideas. Instead, Galison has argued, "what matters is coordination, *not* a full-fledged agreement about signification" (2010, 35; emphasis in the original). This is because "trade focuses on coordinated, local actions enabled by the *thinness* of interpretation rather than the thickness of consensus" (Galison 2010, 36; emphasis in the original). This is a critical point for the perspective we advocate in this book, that is, that the blurring of genres of scholarship on social media sustained by comparative perspectives does not require that these various perspectives come to broad agreements among the various traditions of inquiry involved. All that matters is the shared commitment to guiding epistemological principles that orient scholarship alongside cross-national and regional, cross-media, and cross-platform dynamics. Furthermore, this stance can potentially apply not just to social media but to other objects of inquiry in the field of communication and media studies.

Another important idea regarding trading zones is that "nothing in the notion of trade presupposes some universal notion of a neutral currency" (Galison 1997, 803). Furthermore, "the pertinent theoretical point is that coordination of action occurs between languages in the absence of a full-blown translation" (Galison 1997, 833). The implication of this is that the trading of ideas can proceed without massive investments in shared conceptual frameworks and the potentially ensuing power struggles among traditions of inquiry. On the contrary, the trades happen through the development of contact languages that enable local coordination without the need for global agreements. In his account of the evolution of twentieth-century physics, Galison shows time and again how "it lies among our linguistic abilities to create these mediating contact languages and to do so in a variety of registers" (1997, 833). We suggest that the simple and intuitive vocabulary we have proposed in this book of dimensions of comparison; topics, approaches, methods, and interpretations; and pathways such as languages and histories, provide some initial building blocks to begin developing the contact languages that could assist in trading key ideas among the various traditions of inquiry involved.

A common thread among the reflections about the growth of specialized knowledge in media and communication scholarship has been to underscore the negative effects of this specialization—as connoted, for instance, in the notion of fragmentation. However, Galison has proposed that what accounts for the strength of scientific inquiry is its disunity: "[S]cience is disunified, and—against our intuitions—it is precisely the *dis*unification of science that underpins its strength and stability" (1999, 137; emphasis in the original). Along similar lines, Barbie Zelizer (2016) has recently suggested that dynamics of disunity—plurality, in her framing—not only underpin the field of communication and media studies but are also a basis of its contribution to the larger "fan of disciplines" within the humanities and the social and behavioral sciences. In her view, "Communication's relationship to evidence pushes the fan of disciplines by reminding them of epistemic plurality, or the multiplicity of available interventions" (Zelizer 2016, 227). If Galison and Zelizer, among others, are right about the potentials of intellectual disunity, the need for comparative perspectives as trading zones is greater than ever. This is because what might be at stake is not so much ameliorating the downsides of specialization but fostering the strength and stability of scholarship in the field at large.

To illustrate the relationship between disunity of knowledge on the one hand, and strength and stability of domains of inquiry on the other, Galison (1997) resorts to the analogy of a cable put forward by Charles Sanders Peirce. "Reasoning should not form a chain which is no stronger than the weakest link, but a cable whose fibres may be ever so slender, provided that they are sufficiently numerous and intimately connected," Peirce argued (1984, 213). In this scenario, strength does not emerge from a single unified entity—however powerful it might be—but from the joining of different entities, despite potentially being "slender" or weak. Yet, Galison (1997) cautions: "Ultimately the cable metaphor too takes itself apart, for Peirce insists that the strands not only be 'sufficiently numerous' but also 'intimately connected.' In the cable, that connection is mere physical adjacency, a relation unhelpful in explicating the ties that bind concepts, arguments, instruments, and scientific subcultures. No mechanical analogy will ever be sufficient to do that because it is by coordinating different symbolic and material actions that people create the binding culture of science. All metaphors come to an end" (844).

Against the backdrop of the limitations of mechanical metaphors of that kind, we have emphasized the power of comparative ways of knowing enabled by the actions of blurring boundaries across disparate dimensions of relevant phenomena and traditions of inquiry. It is our hope that the epistemological turn articulated in this book might amount to new beginnings.

Notes

Chapter 1

1. Source: https://datareportal.com/social-media-users.

2. Source: https://en.wikipedia.org/wiki/List_of_social_platforms_with_at_least_100_mil lion_active_users.

3. The popularity of this scholarship has ushered the field into a series of reflections about the underpinnings of this work since the 1990s (Gurevitch and Blumler 1990; Blumler, McLeod, and Rosengren 1992; Esser and Pfetsch 2004; Norris 2009; Esser and Hanitzsch 2012).

Chapter 2

1. Source: https://www.youtube.com/watch?v=u7IgDMOapvo.

2. Source: https://youtu.be/u7IgDMOapvo?t=307.

3. Source: https://www.pagina12.com.ar/294149-time-incluyo-al-colectivo-las-tesis -entre-las-100-personalid.

4. Source: https://diariofemenino.com.ar/mapa-interactivo-muestra-el-impacto-de-un-vio lador-en-tu-camino-en-el-mundo/.

5. Source: https://time.com/collection/100-most-influential-people-2020/5888485/las tesis/.

6. Source: https://www.youtube.com/watch?v=iU2_wg0wYuI.

7. Source: https://www.pagina12.com.ar/101880-mensaje-de-odio-disciplinador.

8. Source: http://dapp.fgv.br/morte-de-marielle-franco-mobiliza-mais-de-567-mil-men coes-no-twitter-aponta-levantamento-da-fgv-dapp/.

9. Sources: https://actualidad.rt.com/actualidad/376220-protestas-rio-janeiro-asesinato -marielle-franco. https://law.utexas.edu/humanrights/projects/marielle-franco-and-the -brazilian-necropolis-assassination-and-after-lives/.

10. Source: https://nyti.ms/2XMtUMa.

11. Source: https://www.bbc.com/news/av/world-52967551.

12. Source: https://www.youtube.com/watch?v=BAETGJl1eec.

13. Source: https://www.standard.co.uk/news/world/blackout-tuesday-shouldnt-use-black livesmatter-hashtag-a4457581.html.

14. Source: https://www.visualcapitalist.com/visualizing-the-social-media-universe-in -2020/.

15. Source: https://www.nytimes.com/2010/12/16/world/europe/16greece.html.

16. Source: https://time.com/4477300/alan-kurdi-photo-one-year-later/.

Chapter 3

1. Source: https://about.fb.com/news/2006/09/facebook-expansion-enables-more-people -to-connect-with-friends-in-a-trusted-environment/.

2. Source: https://www.youtube.com/watch?v=jNQXAC9IVRw.

3. Source: https://www.youtube.com/watch?v=gUy-VCAtHRY.

4. Source: https://nyti.ms/2KPXw5r. Similarly, in 2021, an interviewee from *The New York Times Presents* documentary *Framing Britney Spears* explains that "with the rise of Instagram, it's no longer the tabloids who choose how the world sees Britney."

5. Source: https://www.nytimes.com/2019/03/30/style/kardashians-interview.html.

6. Source: https://en.wikipedia.org/wiki/The_Young_Turks.

7. Source: https://en.wikipedia.org/wiki/Hasan_Piker.

8. Source: https://www.nytimes.com/2020/11/10/style/hasan-piker-twitch.html.

9. Source: https://www.nytimes.com/2020/06/18/technology/protesters-live-stream -twitch.html.

10. Source: https://www.nytimes.com/2020/11/10/style/hasan-piker-twitch.html.

11. Source: https://www.nytimes.com/1960/09/27/archives/tv-the-great-debate-first-nix on-and-kennedy-discussion-is-called-a.html.

12. Source: https://www.nytimes.com/1976/09/23/archives/nixonkennedy-great-debate -gave-a-contrast-in-appearance.html.

13. Source: https://en.wikipedia.org/wiki/Social_media_use_by_Barack_Obama.

14. It would be interesting to reframe this inquiry in light of recent research on the ways in which increasingly professionalized practices of self-branding on social media effectively require third parties (Duffy 2017; Arriagada and Ibáñez 2020).

15. Source: https://en.wikipedia.org/wiki/2008_California_Proposition_8#:~:text=Pro position%208%2C%20known%20informally%20as,was%20later%20overturned%20 in%20court.

Chapter 4

1. Source: https://twitter.com/Twitter/status/1328684389388185600?s=20.

2. Source: https://www.buzzfeed.com/hannahmarder/reactions-to-twitters-fleets-feature.

3. The Instagram account @insta_repeat, which has around 400,000 followers to date, focuses on putting together posts where similar Instagram posts are presented in a grid, almost as if it were an exercise in iconography. QZ magazine wrote an article about it, titled "You are not original or creative on Instagram": https://qz.com/quartzy /1349585/you-are-not-original-or-creative-on-instagram.

4. Source: https://knowyourmeme.com/memes/will-now-have-stories.

5. Source: https://www.nytimes.com/2020/11/17/technology/twitter-fleets-disappear ing-tweets.html.

6. Source: https://twitter.com/kyalbr/status/1328730037789474816?s=20.

7. Source: https://twitter.com/kyalbr/status/1328733178404638720?s=20.

8. Source: https://twitter.com/Twitter/status/1415362679095635970?s=20.

9. Source: https://en.wikipedia.org/wiki/Christchurch_mosque_shootings.

10. Source: https://www.newyorker.com/news/news-desk/inside-the-team-at-facebook -that-dealt-with-the-christchurch-shooting.

11. Source: https://www.lemonde.fr/pixels/article/2019/04/24/paris-lancera-un-appel-de -christchurch-pour-la-suppression-rapide-des-contenus-terroristes-en-ligne_5454218 _4408996.html.

12. Source: https://www.diplomatie.gouv.fr/en/french-foreign-policy/digital-diplomacy /news/article/the-christchurch-call-one-year-on.

13. Source: https://helenclark.foundation/publications-and-media/anti-social-media/.

14. Sources: https://www.vice.com/en/article/43jdbj/christchurch-attack-videos-still-on -facebook-instagram; https://www.lemonde.fr/pixels/article/2019/04/29/moderation-de -la-haine-supremaciste-ce-que-font-ou-pas-facebook-youtube-et-twitter-depuis-christ church_5456454_4408996.html.

15. Source: https://www.theguardian.com/books/2013/nov/19/selfie-word-of-the-year -oed-olinguito-twerk.

16. Source: https://blog.collinsdictionary.com/language-lovers/collins-2017-word-of-the -year-shortlist/.

17. For instance: https://www.smh.com.au/national/in-a-dark-place-adam-goodes-the -nation-and-the-race-question-20150731-giolfa.html.

Chapter 5

1. Source: https://www.washingtonpost.com/news/the-switch/wp/2018/04/11/channel ing-the-social-network-lawmaker-grills-zuckerberg-on-his-notorious-beginnings.

2. Source: https://www.thecrimson.com/article/2003/11/19/facemash-creator-survives -ad-board-the/.

3. Source: https://www.reuters.com/article/us-brazil-banks-payment-platform-idUSK BN27W25C.

4. Source: https://timesofindia.indiatimes.com/tech-news/mark-zuckerberg-changes-his -profile-picture-to-support-digital-india/articleshow/49128369.cms.

5. Source: https://www.youtube.com/watch?v=NKF3PKmTIx0.

6. Source: https://techcrunch.com/2021/02/25/india-announces-sweeping-guidelines-for -social-media-on-demand-streaming-firms-and-digital-news-outlets/.

7. Source: https://video.repubblica.it/sport/addio-a-diego-armando-maradona-da-pele-a -messi-il-mondo-del-calcio/371830/372436.

8. Source: https://www.espn.com/soccer/argentina-arg/story/4245036/diego-maradona -dies-at-the-age-of-60-how-social-media-reacted.

9. Source: https://www.indiatoday.in/television/celebrity/story/sidharth-shukla-to-karan vir-bohra-tv-celebs-mourn-death-of-diego-maradona-1744261-2020-11-26.

10. Source: https://es.wikipedia.org/wiki/Flogger.

Chapter 6

1. Artist Takashi Murakami explained, in an Instagram post devoted to *Yonaguni*, that "Within Japan there are many listeners who are curious about the Japanese lyrics Bad

Bunny sings. Things are heating up here, with blogs popping up exploring questions such as: 'Why Japanese?' and 'Why Yonaguni?'" Source: https://www.instagram.com /p/CSa1zS7nQr6/.

2. Source: https://youtu.be/DhfggDAyUVI.

3. Source: https://www.billboard.com/music/latin/bad-bunny-yhlqmdlg-new-album -interview-9325741/.

4. Source: https://www.vatican.va/content/francesco/es/messages/communications/doc uments/papa-francesco_20190124_messaggio-comunicazioni-sociali.html.

5. Source: https://www.unhcr.org/en-us/news/press/2020/6/5ee9db2e4/1-cent-human ity-displaced-unhcr-global-trends-report.html.

6. Italian for "translator, traitor."

Chapter 7

1. Different historiographies of the field have developed across different national contexts and regions, and across different languages, such as Africa (Willems 2014), Asia (So 2010), Europe (Averbeck 2008), or Latin America (Zarowsky 2017).

References

Abdenour, Jesse. 2017. Digital gumshoes: Investigative journalists' use of social media in television news reporting. *Digital Journalism* 5 (4): 472–492. https://doi.org /10.1080/21670811.2016.1175312.

Abokhodair, Norah, and Adam Hodges. 2019. Toward a transnational model of social media privacy: How young Saudi transnationals do privacy on Facebook. *New Media & Society* 21 (5): 1105–1120. https://doi.org/10.1177/1461444818821363.

Alencar, Amanda. 2018. Refugee integration and social media: A local and experiential perspective. *Information, Communication & Society* 21 (11): 1588–1603. https:// doi.org/10.1080/1369118X.2017.1340500.

Alonso, Andoni, and Pedro Oiarzabal, eds. 2010. *Diasporas in the New Media Age: Identity, Politics, and Community*. Reno: University of Nevada Press.

Anderson, Benedict. 1983. *Imagined Communities: Reflections on the Origin and Spread of Nationalism*. London: Verso Books.

Anderson, Benedict. 1994. Exodus. *Critical Inquiry* 20 (2): 314–327. https://doi.org /10.1086/448713.

Anduiza, Eva, Camilo Cristancho, and José M. Sabucedo. 2014. Mobilization through online social networks: The political protest of the *indignados* in Spain. *Information, Communication & Society* 17 (6): 750–764. https://doi.org/10.1080/1369118X.2013 .808360.

Appadurai, Arjun. 1990. Disjuncture and difference in the global cultural economy. *Theory, Culture & Society* 7 (2–3): 295–310. https://doi.org/10.1177/026327690007 002017.

Appadurai, Arjun. 1996. *Modernity at Large: Cultural Dimensions of Globalization*. Minneapolis: University of Minnesota Press.

Arriagada, Arturo, and Francisco Ibáñez. 2020. "You need at least one picture daily, if not, you're dead": Content creators and platform evolution in the social media ecology. *Social Media + Society* 6 (3): 1–12. https://doi.org/10.1177/2056305120944624.

Arthurs, Jane, Sophia Drakopoulou, and Alessandro Gandini. 2018. Researching YouTube. *Convergence: The International Journal of Research into New Media Technologies* 24 (1): 3–15. https://doi.org/10.1177/1354856517737222.

Averbeck, Stefanie. 2008. Comparative history of communication studies: France and Germany. *The Open Communication Journal* 2 (1): 1–13.

Avital, Moran. 2021. "Days of mourning are days of reconciliation": An analysis of the coverage of the death of controversial Israeli public figures. *Journalism* 22 (7): 1739–1756. https://doi.org/10.1177/1464884918824234.

Bakshy, Eytan, Solomon Messing, and Lada A. Adamic. 2015. Exposure to ideologically diverse news and opinion on Facebook. *Science* (6239): 1130–1132. https://doi.org/10.1126/science.aaa1160.

Barnard, Stephen R. 2016. "Tweet or be sacked": Twitter and the new elements of journalistic practice. *Journalism* 17 (2): 190–207. https://doi.org/10.1177/1464884914553079.

Barton, David, and Carmen Lee. 2013. *Language Online: Investigating Digital Texts and Practices*. Abingdon, UK: Routledge.

Bastos, Marco Toledo, Cornelius Puschmann, and Rodrigo Travitzki. 2013. Tweeting across hashtags: Overlapping users and the importance of language, topics, and politics. In *Proceedings of the 24th ACM Conference on Hypertext and Social Media*, 164–168. New York: Association for Computing Machinery. https://doi.org/10.1145/2481492.2481510.

Bayer, Joseph B., Nicole B. Ellison, Sarita Y. Schoenebeck, and Emily B. Falk. 2016. Sharing the Small moments: Ephemeral social interaction on Snapchat. *Information, Communication & Society* 19 (7): 956–977. https://doi.org/10.1080/1369118X.2015.1084349.

Baym, Nancy K. 2015. *Personal Connections in the Digital Age*. Malden, MA: Polity.

Beck, Ulrich. 2000. *What Is Globalization?* Cambridge: Polity.

Beniger, James R. 1992. Comparison, yes, but—the case of technological and cultural change. In *Comparatively Speaking: Communication and Culture Across Space and Time*, edited by Jay G. Blumler, Jack McLeod, and Karl E. Rosengren, 35–50. Newbury Park, CA: SAGE.

Benjamin, Walter. (1936) 1969. The work of art in the age of mechanical reproduction. In *Illuminations*, edited by Hannah Arendt, 217–251. New York: Schocken.

Benkler, Yochai, Rob Faris, and Hal Roberts. 2018. *Network Propaganda: Manipulation, Disinformation, and Radicalization in American Politics*. New York: Oxford University Press.

Bennett, W. Lance. 2012. The personalization of politics: Political identity, social media, and changing patterns of participation. *The Annals of the American Academy of Political and Social Science* 644 (1): 20–39. https://doi.org/10.1177/0002716 212451428.

Bennett, W. Lance, and Alexandra Segerberg. 2012. The Logic of connective action: Digital media and the personalization of contentious politics. *Information, Communication & Society* 15 (5): 739–768. https://doi.org/10.1080/1369118X.2012 .670661.

Benoit, William L., Mark J. Glantz, Anji L. Phillips, Leslie A. Rill, Corey B. Davis, Jayne R. Henson, and Leigh Anne Sudbrock. 2011. Staying "on message": Consistency in content of presidential primary campaign messages across media. *American Behavioral Scientist* 55 (4): 457–468. https://doi.org/10.1177/0002764211398072.

Benzecry, Claudio E., and Daniel Winchester. 2017. Varieties of microsociology. In *Social Theory Now*, edited by Claudio E. Benzecry, Monika Krause, and Isaac Reed, 42–74. Chicago: University of Chicago Press.

Berger, Charles R. 1991. Communication theories and other curios. *Communication Monographs* 58 (1): 101–113. https://doi.org/10.1080/03637759109376216.

Bhugra, Dinesh. 1993. Cross-cultural aspects of jealousy. *International Review of Psychiatry* 5 (2–3): 271–280. https://doi.org/10.3109/09540269309028317.

Bijker, Wiebe. 1995. *Of Bicycles, Bakelites, and Bulbs: Toward a Theory of Sociotechnical Change*. Cambridge, MA: MIT Press.

Bijker, Wiebe E., Thomas P. Hughes, and Trevor Pinch. 1987. *The Social Construction of Technological Systems: New Directions in the Sociology and History of Technology*. Cambridge, MA: MIT Press.

Billig, Michael. 1995. *Banal Nationalism*. London: SAGE.

Bimber, Bruce. 2014. Digital media in the Obama campaigns of 2008 and 2012: Adaptation to the personalized political communication environment. *Journal of Information Technology & Politics* 11 (2): 130–150. https://doi.org/10.1080/19331681 .2014.895691.

Blumer, Herbert. 1969. *Symbolic Interactionism: Perspective and Method*. Englewood Cliffs, NJ: Prentice-Hall.

Blumler, Jay L., and Michael Gurevitch. 1975. Towards a comparative framework for political communication research. In *Political Communication: Issues and Strategies for Research*, edited by Steven H. Chaffee, 165–193. Beverly Hills, CA: SAGE.

Blumler, Jay G., Jack McLeod, and Karl E. Rosengren, eds. 1992. *Comparatively Speaking: Communication and Culture Across Space and Time*. Newbury Park, CA: SAGE.

Boczkowski, Pablo J. 2004. *Digitizing the News: Innovation in Online Newspapers*. Cambridge, MA: MIT Press.

Boczkowski, Pablo J. 2021. *Abundance: On the Experience of Living in a World of Information Plenty*. New York: Oxford University Press.

Boczkowski, Pablo J., and Leah A. Lievrouw. 2008. Bridging STS and communication studies: Research on media and information technologies. In *The Handbook of Science and Technology Studies*, edited by Edward J. Hackett, Olga Amsterdamska, Michael E. Lynch, Judy Wajcman, Sergio Sismondo, Wiebe E. Bijker, Stephen Turner, et al., 949–977. Cambridge, MA: MIT Press.

Boczkowski, Pablo J., Mora Matassi, and Eugenia Mitchelstein. 2018. How young users deal with multiple platforms: The role of meaning-making in social media repertoires. *Journal of Computer-Mediated Communication* 23 (5): 245–259. https://doi.org/10.1093/jcmc/zmy012.

Boczkowski, Pablo, and Eugenia Mitchelstein. 2021. *The Digital Environment: How We Live, Learn, Work, and Play Now*. Cambridge, MA: MIT Press.

Boczkowski, Pablo J., Eugenia Mitchelstein, and Martin Walter. 2011. Convergence across divergence: Understanding the gap in the online news choices of journalists and consumers in Western Europe and Latin America. *Communication Research* 38 (3): 376–396. https://doi.org/10.1177/0093650210384989.

Bode, Leticia. 2016. Pruning the news feed: Unfriending and unfollowing political content on social media. *Research & Politics* 3 (3): 1–8. https://doi.org/10.1177/2053168016661873.

Bode, Leticia, and Emily K. Vraga. 2018. Studying politics across media. *Political Communication* 35 (1): 1–7. https://doi.org/10.1080/10584609.2017.1334730.

Bogers, Loes, Sabine Niederer, Federica Bardelli, and Carlo De Gaetano. 2020. Confronting bias in the online representation of pregnancy. *Convergence* 26 (5–6): 1037–1059. https://doi.org/10.1177/1354856520938606.

Bolter, J. David, and Richard A. Grusin. 1999. *Remediation: Understanding New Media*. Cambridge, MA: MIT Press.

Bondes, Maria, and Günter Schucher. 2014. Derailed emotions: The transformation of claims and targets during the Wenzhou online incident. *Information, Communication & Society* 17 (1): 45–65. https://doi.org/10.1080/1369118X.2013.853819.

Bosch, Tanja Estella, Mare Admire, and Meli Ncube. 2020. Facebook and politics in Africa: Zimbabwe and Kenya. *Media, Culture & Society* 42 (3): 349–364. https://doi.org/10.1177/0163443719895194.

Bou-Franch, Patricia, and Pilar Garcés-Conejos Blitvich, eds. 2019. *Analyzing Digital Discourse: New Insights and Future Directions*. Cham, Switzerland: Palgrave Macmillan.

Boulianne, Shelley. 2015. Social media use and participation: A meta-analysis of current research. *Information, Communication & Society* 18 (5): 524–538. https://doi.org/10.1080/1369118X.2015.1008542.

Boulianne, Shelley. 2020. Twenty years of digital media effects on civic and political participation. *Communication Research* 47 (7): 947–966. https://doi.org/10.1177/0093650218808186.

Bourdon, Jérôme. 2018. The case for the technological comparison in communication history. *Communication Theory* 28 (1): 89–109. https://doi.org/10.1093/ct/qtx001.

Bouvier, Gwen. 2019. How journalists source trending social media feeds: A critical discourse perspective on Twitter. *Journalism Studies* 20 (2): 212–231. https://doi.org/10.1080/1461670X.2017.1365618.

boyd, danah. 2014. *It's Complicated: The Social Lives of Networked Teens*. New Haven, CT: Yale University Press.

boyd, danah, and Kate Crawford. 2012. Critical questions for big data: Provocations for a cultural, technological, and scholarly phenomenon. *Information, Communication & Society* 15 (5): 662–679. https://doi.org/10.1080/1369118X.2012.678878.

boyd, danah, and Nicole B. Ellison. 2007. Social network sites: Definition, history and scholarship. *Journal of Computer-Mediated Communication* 13 (1): 210–230. https://doi.org/10.1111/j.1083-6101.2007.00393.x.

Bozdag, Cigdem. 2020. Managing diverse online networks in the context of polarization: Understanding how we grow apart on and through social media. *Social Media + Society* 6 (4): 1–13. https://doi.org/10.1177/2056305120975713.

Bozdag, Cigdem, and Kevin Smets. 2017. Understanding the images of Alan Kurdi with "small data": A qualitative, comparative analysis of tweets about refugees in Turkey and Flanders (Belgium). *International Journal of Communication* 11:4046–4069.

Brems, Cara, Martina Temmerman, Todd Graham, and Marcel Broersma. 2017. Personal branding on Twitter: How employed and freelance journalists stage themselves on social media. *Digital Journalism* 5 (4): 443–459. https://doi.org/10.1080/21670811.2016.1176534.

Brinkerhoff, Jennifer M. 2009. *Digital Diasporas: Identity and Transnational Engagement*. New York: Cambridge University Press.

Brock, André L. 2020. *Distributed Blackness: African American Cybercultures*. New York: New York University Press.

Broniatowski, David A., Amelia M. Jamison, SiHua Qi, Lulwah AlKulaib, Tao Chen, Adrian Benton, Sandra C. Quinn, and Mark Dredze. 2018. Weaponized health communication: Twitter bots and Russian trolls amplify the vaccine debate. *American Journal of Public Health* 108 (10): 1378–1384. https://doi.org/10.2105/AJPH.2018 .304567.

Brosius, Hans-Bernd, and Hans Mathias Kepplinger. 1990. The agenda-setting function of television news: Static and dynamic views. *Communication Research* 17 (2): 183–211. https://doi.org/10.1177/009365090017002003.

Brüggemann, Michael, Sven Engesser, Florin Büchel, Edda Humprecht, and Laia Castro. 2014. Hallin and Mancini revisited: Four empirical types of Western media systems. *Journal of Communication* 64 (6): 1037–1065. https://doi.org/10.1111/jcom .12127.

Bruns, Axel. 2019. *Are Filter Bubbles Real?* Cambridge: Polity.

Brydges, Taylor, and Jenny Sjöholm. 2019. Becoming a personal style blogger: Changing configurations and spatialities of aesthetic labour in the fashion industry. *International Journal of Cultural Studies* 22 (1): 119–139. https://doi.org/10.1177 /1367877917752404.

Bucher, Taina. 2012. Want to be on the top? Algorithmic power and the threat of invisibility on Facebook. *New Media & Society* 14 (7): 1164–1180. https://doi.org/10 .1177/1461444812440159.

Bucher, Taina. 2021. *Facebook*. Cambridge: Polity.

Burgess, Jean, and Nancy K. Baym. 2020. *Twitter: A Biography*. New York: New York University Press.

Burgess, Jean, and Joshua Green. 2018. *YouTube: Online Video and Participatory Culture*. Cambridge: Polity.

Burgess, Jean, Alice E. Marwick, and Thomas Poell. 2018. *The SAGE Handbook of Social Media*. London: SAGE Reference.

Canter, Lily. 2015. Personalised tweeting: The emerging practices of journalists on Twitter. *Digital Journalism* 3 (6): 888–907. https://doi.org/10.1080/21670811.2014 .973148.

Castañeda, Ernesto. 2012. The *Indignados* of Spain: A precedent to Occupy Wall Street. *Social Movement Studies* 11 (3–4): 309–319. https://doi.org/10.1080/14742837 .2012.708830.

Castells, Manuel. 2004. *The Network Society: A Cross-Cultural Perspective*. Cheltenham, UK: Edward Elgar.

Chadwick, Andrew. 2017. *The Hybrid Media System: Politics and Power*. New York: Oxford University Press.

Chan, Joseph M., and Chin-Chuan Lee. 1984. Journalistic "paradigms" of civil protests: A case study in Hong Kong. In *The News Media in National and International Conflict*, edited by Andrew Arno, and Wimal Dissanayake, 183–202. Boulder, CO: Westview Press.

Chen, Hsuan-Ting, Michael Chan, and Francis L. F. Lee. 2016. Social media use and democratic engagement: A comparative study of Hong Kong, Taiwan, and China. *Chinese Journal of Communication* 9 (4): 348–366. https://doi.org/10.1080/17544750.2016 .1210182.

Cheruiyot, David. 2021. The (other) anglophone problem: Charting the development of a journalism subfield. *African Journalism Studies* 42 (2): 94–105. https://doi .org/10.1080/23743670.2021.1939750.

Chorianopoulos, Konstantinos, and George Lekakos. 2008. Introduction to social TV: Enhancing the shared experience with interactive TV. *International Journal of Human–Computer Interaction* 24 (2): 113–120. https://doi.org/10.1080/10447310701821574.

Christin, Angèle, and Rebecca Lewis. 2021. The drama of metrics: Status, spectacle, and resistance among YouTube drama creators. *Social Media + Society* 7 (1): 1–14. https://doi.org/10.1177/2056305121999660.

Chu, Shu-Chuan, and Sejung Marina Choi. 2010. Social capital and self-presentation on social networking sites: A comparative study of Chinese and American young generations. *Chinese Journal of Communication* 3 (4): 402–420. https://doi.org/10.1080 /17544750.2010.516575.

Chua, Trudy Hui, and Leanne Chang. 2016. Follow me and like my beautiful selfies: Singapore teenage girls' engagement in self-presentation and peer comparison on social media. *Computers in Human Behavior* 55:190–197. https://doi.org/10.1016/j .chb.2015.09.011.

Chun, Wendy Hui Kyong. 2008. *Updating to Remain the Same: Habitual New Media*. Cambridge, MA: MIT Press.

Chun, Wendy Hui Kyong. 2011. The enduring ephemeral, or the future is a memory. In *Media Archaeology: Approaches, Applications, and Implications*, edited by Erkki Huhtamo and Jussi Parikka, 184–206. Berkeley: University of California Press.

Chun, Wendy Hui Kyong, Anna Watkins Fisher, and Thomas Keenan, eds. 2005. *New Media, Old Media: A History and Theory Reader*. New York: Routledge.

Cobo, Leila. 2020. Bad Bunny talks surprise new album "El ultimo tour del mundo" & Rosalia collab. *Billboard*, November, 27, 2020. https://www.billboard.com/articles /columns/latin/9490254/bad-bunny-surprise-album-interview-el-ultimo-tour-del -mundo-rosalia.

Cohen, Bernard Cecil. 1963. *The Press and Foreign Policy*. Princeton, NJ: Princeton University Press.

Colleoni, Elanor, Alessandro Rozza, and Adam Arvidsson. 2014. Echo chamber or public sphere? Predicting political orientation and measuring political homophily in Twitter using big data. *Journal of Communication* 64 (2): 317–332. https://doi.org /10.1111/jcom.12084.

Collier, David. 1993. The comparative method. In *Political Science: The State of Discipline II*, edited by Ada W. Finifter, 105–119. Washington, DC: American Political Science Association.

Corazza, Michele, Stefano Menini, Elena Cabrio, Sara Tonelli, and Serena Villata. 2019. Cross-platform evaluation for Italian hate speech detection. In *CLiC-it 2019– 6th Annual Conference of the Italian Association for Computational Linguistics*. Bari, Italy.

Costanza-Chock, Sasha. 2014. *Out of the Shadows, Into the Streets!: Transmedia Organizing and the Immigrant Rights Movement*. Cambridge, MA: MIT Press.

Craig, David, Jian Lin, and Stuart Cunningham. 2021. *Wanghong as Social Media Entertainment in China*. Cham, Switzerland: Palgrave Macmillan.

Crawford, Kate, and Tarleton Gillespie. 2016. What is a flag for? Social media reporting tools and the vocabulary of complaint. *New Media & Society* 18 (3): 410–428. https://doi.org/10.1177/1461444814543163.

Curran, James, and Myung-Jin Park, eds. 2000. *De-Westernizing Media Studies*. London: Routledge.

Czitrom, Daniel J. 1982. *Media and the American Mind: From Morse to McLuhan*. Chapel Hill: University of North Carolina Press.

deCordova, Richard. 1990. *Picture Personalities: The Emergence of the Star System in America*. Chicago: University of Illinois Press.

de Lenne, Orpha, Laura Vandenbosch, Steven Eggermont, Kathrin Karsay, and Jolien Trekels. 2020. Picture-perfect lives on social media: A cross-national study on the role of media ideals in adolescent well-being. *Media Psychology* 23 (1): 52–78. https://doi .org/10.1080/15213269.2018.1554494.

Del Vicario, Michela, Gianna Vivaldo, Alessandro Bessi, Fabiana Zollo, Antonio Scala, Guido Caldarelli, and Walter Quattrociocchi. 2016. Echo chambers: Emotional contagion and group polarization on Facebook. *Scientific Reports* 6:1–12. https://www .nature.com/articles/srep37825.

de Saussure, Ferdinand. (1916) 1983. *Course in General Linguistics*, translated and annotated by Roy Harris. London: Duckworth.

Desjardins, Renée. 2016. *Translation and Social Media: In Theory, in Training and in Professional Practice*. London: Palgrave.

de Sola Pool, Ithiel. 1983. *Technologies of Freedom*. Cambridge, MA: Belknap Press.

Deuze, Mark. 2003. The web and its journalisms: Considering the consequences of different types of newsmedia online. *New Media & Society* 5 (2): 203–230. https://doi .org/10.1177/1461444803005002004.

DeVito, Michael Ann, Jeremy Birnholtz, and Jeffery T. Hancock. 2017. Platforms, people, and perception: Using affordances to understand self-presentation on social media. In *Proceedings of the 2017 ACM Conference on Computer Supported Cooperative Work and Social Computing*, 740–754. New York: Association for Computing Machinery. https://doi.org/10.1145/2998181.2998192.

DeVito, Michael Ann, Ashley Marie Walker, and Jeremy Birnholtz. 2018. "Too gay for Facebook": Presenting LGBTQ+ identity throughout the personal social media ecosystem. In *Proceedings of the ACM on Human-Computer Interaction CSCW*, 1–23. New York: Association for Computing Machinery. https://dl.acm.org/doi/10.1145/3274313.

Dimmick, John W. 2003. *Media Competition and Coexistence: The Theory of the Niche*. Mahwah, NJ: Lawrence Erlbaum Associates.

Dimmick, John, Yan Chen, and Zhan Li. 2004. Competition between the Internet and traditional news media: The gratification-opportunities niche dimension. *The Journal of Media Economics* 17 (1): 19–33. https://doi.org/10.1207/s15327736me1701_2.

Dimmick, John, John Christian Feaster, and Artemio Ramirez Jr. 2011. The niches of interpersonal media: Relationships in time and space. *New Media & Society* 13 (8): 1265–1282. https://doi.org/10.1177/1461444811403445.

Donath, Judith, and danah boyd. 2004. Public displays of connection. *BT Technology Journal* 22 (4): 71–82. https://doi.org/10.1023/B:BTTJ.0000047585.06264.cc

Doughty, Mark, Duncan Rowland, and Shaun Lawson. 2012. Who is on your sofa? TV audience communities and second screening social networks. In *Proceedings of the 10th European Conference on Interactive TV and Video*, 79–86. New York: Association for Computing Machinery.

Douglas, Susan J. 1989. *Inventing American Broadcasting, 1899–1922*. Baltimore: Johns Hopkins University Press.

Douglas, Susan J., and Andrea M. McDonnell. 2019. *Celebrity: A History of Fame*. New York: New York University Press.

Druckman, James N. 2003. The power of television images: The first Kennedy-Nixon debate revisited. *The Journal of Politics* 65 (2): 559–571. https://doi.org/10.1111/1468 -2508.t01-1-00015.

Dubois, Elizabeth, and Grant Blank. 2018. The echo chamber is overstated: The moderating effect of political interest and diverse media. *Information, Communication & Society* 21 (5): 729–745. https://doi.org/10.1080/1369118X.2018.1428656.

Dubrofsky, Rachel E. 2011. Surveillance on reality television and Facebook: From authenticity to flowing data. *Communication Theory* 21 (2): 111–129. https://doi.org /10.1111/j.1468-2885.2011.01378.x.

Duffy, Brooke Erin. 2017. *(Not) Getting Paid to Do What You Love: Gender, Social Media, and Aspirational Work*. New Haven, CT: Yale University Press.

Duffy, Brooke Erin, and Ngai Keung Chan. 2019. "You never really know who's looking": Imagined surveillance across social media platforms. *New Media & Society* 21 (1): 119–138. https://doi.org/10.1177/1461444818791318.

Duguay, Stefanie. 2016. Lesbian, gay, bisexual, trans, and queer visibility through selfies: Comparing platform mediators across Ruby Rose's Instagram and Vine presence. *Social Media + Society* 2 (2): 1–12. https://doi.org/10.1177/2056305116641975.

Dwyer, Catherine, Starr Hiltz, and Katia Passerini. 2007. Trust and privacy concern within social networking sites: A comparison of Facebook and MySpace. In *AMCIS 2007 Proceedings*. Keystone, CO: Association for Information Systems (AIS). https:// aisel.aisnet.org/amcis2007/339.

Ellison, Nicole B., Charles Steinfield, and Cliff Lampe. 2007. The benefits of Facebook "friends": Social capital and college students' use of online social network sites. *Journal of Computer-Mediated Communication* 12 (4): 1143–1168. https://doi.org /10.1111/j.1083-6101.2007.00367.x.

Emery, Edwin, and Michael C. Emery. 1978. *The Press and America: An Interpretative History of the Mass Media*, 4th ed. Englewood Cliffs, NJ: Prentice-Hall.

Entman, Robert M., and Nikki Usher. 2018. Framing in a fractured democracy: Impacts of digital technology on ideology, power and cascading network activation. *Journal of Communication* 68 (2): 298–308. https://doi.org/10.1093/joc/jqx019.

Esser, Frank. 2019. Advances in comparative political communication research through contextualization and cumulation of evidence. *Political Communication* 36 (4): 680–686. https://doi.org/10.1080/10584609.2019.1670904.

Esser, Frank, and Thomas Hanitzsch. 2012. *Handbook of Comparative Communication Research*. New York: Routledge.

Esser, Frank, and Barbara Pfetsch, eds. 2004. *Comparing Political Communication: Theories, Cases, and Challenges*. Cambridge: Cambridge University Press.

Esser, Frank, and Rens Vliegenthart. 2017. Comparative research methods. In *The International Encyclopedia of Communication Research Methods*, edited by Jörg Matthes, Christine S. Davis, and Robert F. Potter, 1–22. Hoboken, NJ: Wiley.

Evans, Elizabeth. 2011. *Transmedia Television: Audiences, New Media, and Daily Life.* New York: Routledge.

Faist, Thomas. 2004. The transnational turn in migration research: Perspectives for the study of politics and polity. In *Transnational Spaces: Disciplinary Perspectives*, edited by Maja Povrzanović Frykman, 11–45. Malmö: Malmö University Press.

Ferguson, Douglas A. 1992. Profile: Channel repertoire in the presence of remote control devices, VCRs and cable television. *Journal of Broadcasting & Electronic Media* 36 (1): 83–91. https://doi.org/10.1080/08838159209364156.

Ferguson, Douglas A., and Elizabeth M. Perse. 1993. Media and audience influences on channel repertoire. *Journal of Broadcasting & Electronic Media* 37 (1): 31–47. https://doi.org/10.1080/08838159309364202.

Fischer, Claude S. 1992. *America Calling: A Social History of the Telephone to 1940.* Berkeley: University of California Press.

Flaxman, Seth, Sharad Goel, and Justin M. Rao. 2016. Filter bubbles, echo chambers, and online news consumption. *Public Opinion Quarterly* 80 (S1): 298–320. https://doi.org/10.1093/poq/nfw006.

Flesher Fominaya, Cristina. 2015. Debunking spontaneity: Spain's 15-M/*Indignados* as autonomous movement. *Social Movement Studies* 14 (2): 142–163. https://doi.org/10.1080/14742837.2014.945075.

Fletcher, Richard, Alessio Cornia, and Rasmus Kleis Nielsen. 2020. How polarized are online and offline news audiences? A comparative analysis of twelve countries. *The International Journal of Press/Politics* 25 (2): 169–195. https://doi.org/10.1177/1940161219892768.

Foot, Kirsten. 2014. The online emergence of pushback on social media in the United States: A historical discourse analysis. *International Journal of Communication* 8:1313–1342.

French, Megan, and Natalya N. Bazarova. 2017. Is anybody out there? Understanding masspersonal communication through expectations for response across social media platforms. *Journal of Computer-Mediated Communication* 22 (6): 303–319. https://doi.org/10.1111/jcc4.12197.

Fuchs, Christian. 2016. Baidu, Weibo and Renren: The global political economy of social media in China. *Asian Journal of Communication* 26 (1): 14–41. https://doi.org/10.1080/01292986.2015.1041537.

Fuchs, Christian, and Jack Linchuan Qiu. 2018. Ferments in the field: Introductory reflections on the past, present and future of communication studies. *Journal of Communication* 68 (2): 219–232. https://doi.org/10.1093/joc/jqy008.

Furedi, Frank. 2016. Moral panic and reading: Early elite anxieties about the media effect. *Cultural Sociology* 10 (4): 523–537. https://doi.org/10.1177/1749975515626953.

Gabore, Samuel Mochona, and Deng Xiujun. 2018. Opinion formation in social media: The influence of online news dissemination on Facebook posts. *Communicatio* 44 (2): 20–40. https://doi.org/10.1080/02500167.2018.1504097.

Gainous, Jason, Kevin M. Wagner, and Jason P. Abbott. 2015. Civic disobedience: Does internet use stimulate political unrest in East Asia? *Journal of Information Technology & Politics* 12 (2): 219–236. https://doi.org/10.1080/19331681.2015.1034909.

Galison, Peter. 1997. *Image and Logic.* Chicago: University of Chicago Press.

Galison, Peter. 1999. Trading Zone: Coordinating action and belief (1998 abridgment). In *The Science Studies Reader*, edited by Mario Biagioli, 137–160. New York: Routledge.

Galison, Peter. 2010. Trading with the enemy. In *Trading Zones and International Expertise: Creating New Kinds of Collaboration*, edited by Michael E. Gorman, 25–52. Cambridge, MA: MIT Press.

Gamson, Joshua. 1994. *Claims to Fame: Celebrity in Contemporary America.* Berkeley: University of California Press.

Geertz, Clifford. 1980. Blurred genres: The refiguration of social thought. *The American Scholar* 49 (2): 165–179.

Gerber, Alan S., Dean Karlan, and Daniel Bergan. 2009. Does the media matter? A field experiment measuring the effect of newspapers on voting behavior and political opinions. *American Economic Journal: Applied Economics* 1 (2): 35–52. https://www.aeaweb.org/articles?id=10.1257/app.1.2.35.

Gerbner, George, and Marsha Siefert, eds. 1983. *Ferment in the Field: Communications Scholars Address Critical Issues and Research Tasks of the Discipline.* Philadelphia: Annenberg School Press.

Gibbs, Martin, James Meese, Michael Arnold, Bjorn Nansen, and Marcus Carter. 2015. #Funeral and Instagram: Death, social media, and platform vernacular. *Information, Communication & Society* 18 (3): 255–268. https://doi.org/10.1080/1369118X.2014.987152.

Giglietto, Fabio, and Donatella Selva. 2014. Second screen and participation: A content analysis on a full season dataset of tweets. *Journal of Communication* 64 (2): 260–277. https://doi.org/10.1111/jcom.12085.

Gil de Zúñiga, Homero, and James H. Liu. 2017. Second screening politics in the social media sphere: Advancing research on dual screen use in political communication with evidence from 20 countries. *Journal of Broadcasting & Electronic Media* 61 (2): 193–219. https://doi.org/10.1080/08838151.2017.1309420.

Gil de Zúñiga, Homero, Víctor García-Perdomo, and Shannon C. McGregor. 2015. What is second screening? Exploring motivations of second screen use and its effect on online political participation. *Journal of Communication* 65 (5): 793–815. https://doi.org/10.1111/jcom.12174.

Gil de Zúñiga, Homero, Nakwon Jung, and Sebastián Valenzuela. 2012. Social media use for news and individuals' social capital, civic engagement and political participation. *Journal of Computer-Mediated Communication* 17 (3): 319–336. https://doi.org/10.1111/j.1083-6101.2012.01574.x.

Gillespie, Marie, Souad Osseiran, and Margie Cheesman. 2018. Syrian refugees and the digital passage to Europe: Smartphone infrastructures and affordances. *Social Media + Society* 4 (1): 1–12. https://doi.org/10.1177/2056305118764440.

Gillespie, Tarleton. 2010. The politics of "platforms." *New Media & Society* 12 (3): 347–364. https://doi.org/10.1177/1461444809342738.

Gillespie, Tarleton. 2018. *Custodians of the Internet: Platforms, Content Moderation, and the Hidden Decisions That Shape Social Media.* New Haven, CT: Yale University Press.

Gitelman, Lisa. 2006. *Always Already New: Media, History and the Data of Culture.* Cambridge, MA: MIT Press.

Gitlin, Todd. 1980. *The Whole World Is Watching: Mass Media in the Making & Unmaking of the New Left.* Berkeley: University of California Press.

Glynn, Kevin. 2000. *Tabloid Culture: Trash Taste, Popular Power, and the Transformation of American Television.* Durham, NC: Duke University Press.

Goffman, Erving. 1959. *The Presentation of Self in Everyday Life.* New York: Anchor.

Goffman, Erving. 1967. *Interaction Ritual: Essays in Face to Face Behavior.* Chicago: Aldine Transaction.

Goggin, Gerard, and McLelland, Mark J. 2009. Internationalizing Internet studies: Beyond anglophone paradigms. In *Internationalizing Internet Studies: Beyond Anglophone Paradigms*, edited by Gerard Goggin and Mark J. McLelland, 3–17. New York: Routledge.

Gómez-Cruz, Edgar, and Ignacio Siles. 2021. Visual communication in practice: A texto-material approach to WhatsApp in Mexico City. *International Journal of Communication* 15:4546–4566.

Gottlieb, Nanette. 2009. Language on the internet in Japan. In *Internationalizing Internet Studies: Beyond Anglophone Paradigms*, edited by Gerard Goggin and Mark J. McLelland, 65–78. New York: Routledge.

Gray, Kishonna L. 2020. *Intersectional Tech: Black Users in Digital Gaming.* Baton Rouge, LA: Louisiana State University Press.

Grint, Keith, and Steve Woolgar. 1992. Computers, guns, and roses: What's social about being shot? *Science, Technology, & Human Values* 17 (3): 366–380. https://doi.org /10.1177/016224399201700306.

Guerrero, Manuel, and Mireya Márquez-Ramírez, eds. 2014. *Media Systems and Communication Policies in Latin America*. Houndmills, UK: Palgrave Macmillan.

Gurevitch, Michael, and Jay G. Blumler. 1990. Political communication systems and democratic values. In *Democracy and the Mass Media: A Collection of Essays*, edited by Judith Lichtenberg, 269–289. Cambridge: Cambridge University Press.

Gutiérrez-Martín, Alfonso, and Alba Torrego-González. 2018. The Twitter games: Media education, popular culture and multiscreen viewing in virtual concourses. *Information, Communication & Society* 21 (3): 434–447. https://doi.org/10.1080/1369118X .2017.1284881.

Ha, Louisa, and Ling Fang. 2012. Internet experience and time displacement of traditional news media use: An application of the theory of the niche. *Telematics and Informatics* 29 (2): 177–186. https://doi.org/10.1016/j.tele.2011.06.001.

Hall, Edward T. 1976. *Beyond Culture*. Garden City, NY: Anchor Books.

Hall, Margeret, Athanasios Mazarakis, Martin Chorley, and Simon Caton. 2018. Editorial of the special issue on following user pathways: Key contributions and future directions in cross-platform social media research. *International Journal of Human-Computer Interaction* 34 (10): 895–912. https://doi.org/10.1080/10447318.2018 .1471575.

Hall, Stuart. 1980. Cultural studies: Two paradigms. *Media, Culture & Society* 2 (1): 57–72. https://doi.org/10.1177/016344378000200106.

Hallin, Daniel C., and Paolo Mancini. 2004. *Comparing Media Systems: Three Models of Media and Politics*. New York: Cambridge University Press.

Hallin, Daniel C., and Paolo Mancini. 2017. Ten years after *Comparing Media Systems*: What have we learned? *Political Communication* 34 (2): 155–171. https://doi .org/10.1080/10584609.2016.1233158.

Halpern, Daniel, and Jennifer Gibbs. 2013. Social media as a catalyst for online deliberation? Exploring the affordances of Facebook and YouTube for political expression. *Computers in Human Behavior* 29 (3): 1159–1168. https://doi.org/10.1016 /j.chb.2012.10.008.

Harder, Raymond A., Julie Sevenans, and Peter Van Aelst. 2017. Intermedia agenda setting in the social media age: How traditional players dominate the news agenda in election times. *The International Journal of Press/Politics* 22 (3): 275–293. https://doi.org /10.1177/1940161217704969.

Hargittai, Eszter. 2007. Whose space? Differences among users and non-users of social network sites. *Journal of Computer-Mediated Communication* 13 (1): 276–297. https://doi.org/10.1111/j.1083-6101.2007.00396.x.

Hargittai, Eszter. 2020. Potential biases in big data: Omitted voices on social media. *Social Science Computer Review* 38 (1): 10–24. https://doi.org/10.1177/0894439318788322.

Hargittai, Eszter, and Yu-li Patrick Hsieh. 2010. Predictors and consequences of differentiated practices on social network sites. *Information, Communication & Society* 13 (4): 515–536. https://doi.org/10.1080/13691181003639866.

Harlow, Summer. 2019. Framing #Ferguson: A comparative analysis of media tweets in the U.S., U.K., Spain, and France. *International Communication Gazette* 81 (6–8): 623–643. https://doi.org/10.1177/1748048518822610.

Harlow, Summer, and Thomas J. Johnson. 2011. The Arab Spring| Overthrowing the Protest Paradigm? How *The New York Times*, Global Voices and Twitter Covered the Egyptian Revolution. *International Journal of Communication* 5:1359–1374.

Harp, Dustin, Ingrid Bachmann, and Lei Guo. 2012. The whole online world is watching: Profiling social networking sites and activists in China, Latin America and the United States. *International Journal of Communication* 6:298–321.

Harrington, Stephen, Tim Highfield, and Axel Bruns. 2013. More than a backchannel: Twitter and television. *Participations* 10 (1): 405–408.

Hartley, John. 2018. Pushing back: Social media as an evolutionary phenomenon. In *The SAGE Handbook of Social Media*, edited by Jean Burgess, Alice E. Marwick, and Thomas Poell, 13–33. Los Angeles: SAGE.

Hasebrink, Uwe, and Andreas Hepp. 2017. How to research cross-media practices? Investigating media repertoires and media ensembles. *Convergence* 23 (4): 362–377. https://doi.org/10.1177/1354856517700384.

Hasebrink, Uwe, and Jutta Popp. 2006. Media repertoires as a result of selective media use: A conceptual approach to the analysis of patterns of exposure. *Communications* 31 (3): 369–387. https://doi.org/10.1515/COMMUN.2006.023.

Hayles, N. Katherine. 2007. Intermediation: The pursuit of a vision. *New Literary History* 38 (1): 99–125. https://www.jstor.org/stable/20057991.

Hedman, Ulrika, and Monika Djerf-Pierre. 2013. The social journalist: Embracing the social media life or creating a new digital divide? *Digital Journalism* 1 (3): 368–385. https://doi.org/10.1080/21670811.2013.776804.

Heeter, Carrie. 1985. Program selection with abundance of choice: A process model. *Human Communication Research* 12 (1): 126–152. https://doi.org/10.1111/j.1468-2958.1985.tb00070.x.

Hegde, Radha Sarma. 2016. *Mediating Migration*. Cambridge: Polity.

Hermida, Alfred. 2010. Twittering the news: The emergence of ambient journalism. *Journalism Practice* 4 (3): 297–308. https://doi.org/10.1080/17512781003640703.

Hermida, Alfred. 2014. *Tell Everyone: Why We Share and Why It Matters*. Toronto: Anchor Canada.

Herring, Susan C., ed. 1996. *Computer-Mediated Communication: Linguistic, Social, and Cross-Cultural Perspectives*. Philadelphia: John Benjamins.

Hessel, Stéphane. 2011. *Time for Outrage: Indignez-vous!* New York: Twelve Books.

Hesselberth, Pepita. 2018. Discourses on disconnectivity and the right to disconnect. *New Media & Society* 20 (5): 1994–2010. https://doi.org/10.1177/1461444817711449.

Highfield, Tim, Stephen Harrington, and Axel Bruns. 2013. Twitter as a technology for audiencing and fandom: The #Eurovision phenomenon. *Information, Communication & Society* 16 (3): 315–339.

Highfield, Tim, and Tama Leaver. 2016. Instagrammatics and digital methods: Studying visual social media, from selfies and GIFs to memes and emoji. *Communication Research and Practice* 2 (1): 47–62. https://doi.org/10.1080/22041451.2016.1155332.

Hill, Annette. 2011. *Paranormal Media: Audiences, Spirits and Magic in Popular Culture*. London: Routledge.

Himelboim, Itai, Stephen McCreery, and Marc Smith. 2013. Birds of a feather tweet together: Integrating network and content analyses to examine cross-ideology exposure on Twitter. *Journal of Computer-Mediated Communication* 18 (2): 154–174. https://doi.org/10.1111/jcc4.12001.

Hofstede, Geert H. 1983. National cultures in four dimensions: A research-based theory of cultural differences among nations. *International Studies of Management & Organization* 13 (1–2): 46–74. https://doi.org/10.1080/00208825.1983.11656358.

Hofstede, Geert H. 1984. *Culture's Consequences: International Differences in Work-Related Values*. Abridged ed. Beverly Hills, CA: Sage Publications.

Hofstede, Geert H. 1991. *Cultures and Organizations: Software of the Mind*. London: McGraw-Hill.

Hofstede, Geert H. 1998. Attitudes, values and organizational culture: Disentangling the concepts. *Organization Studies* 19 (3): 477–493. https://doi.org/10.1177/017084069801900305.

Holmes, Su, and Deborah Jermyn. 2004. *Understanding Reality Television*. London: Routledge.

Hopke, Jill E. 2015. Hashtagging politics: Transnational anti-fracking movement Twitter practices. *Social Media + Society* 1 (2): 1–12. https://doi.org/10.1177/2056305115605521.

Horvát, Emőke-Ágnes, and Eszter Hargittai. 2021. Birds of a feather flock together online: Digital inequality in social media repertoires. *Social Media + Society* 7 (4): 1–14. https://doi.org/10.1177/20563051211052897.

Humphreys, Lee. 2018. *The Qualified Self: Social Media and the Accounting of Everyday Life*. Cambridge, MA: MIT Press.

Humphreys, Lee, Phillipa Gill, Balachander Krishnamurthy, and Elizabeth Newbury. 2013. Historicizing new media: A content analysis of Twitter. *Journal of Communication* 63 (3): 413–431. https://doi.org/10.1111/jcom.12030.

Hunt, Melissa G., Rachel Marx, Courtney Lipson, and Jordyn Young. 2018. No more FOMO: Limiting social media decreases loneliness and depression. *Journal of Social and Clinical Psychology* 37 (10): 751–768. https://doi.org/10.1521/jscp.2018.37.10.751.

Innis, Harold A. 1964. *The Bias of Communication*. Toronto: University of Toronto Press.

Ito, Mizuko, Daisuke Okabe, and Misa Matsuda. 2005. *Personal, Portable, Pedestrian: Mobile Phones in Japanese Life*. Cambridge, MA: MIT Press.

Jackson, Linda A., and Jin-Liang Wang. 2013. Cultural differences in social networking site use: A comparative study of China and the United States. *Computers in Human Behavior* 29 (3): 910–921. https://doi.org/10.1016/j.chb.2012.11.024.

Jackson, Sarah J., Moya Bailey, and Brooke Foucault Welles. 2020. *#HashtagActivism: Networks of Race and Gender Justice*. Cambridge, MA: MIT Press.

Jackson, Sarah J., and Brooke Foucault Welles. 2015. Hijacking #MYNYPD: Social media dissent and networked counterpublics. *Journal of Communication* 65 (6): 932–952. https://doi.org/10.1111/jcom.12185.

Jaramillo, Mary Correa. 2006. Desinformación y propaganda: Estrategias de gestión de la comunicación en el conflicto armado colombiano. *Reflexión Política* 8 (15): 94–106.

Jasanoff, Sheila, and Sang-Hyun Kim, eds. 2015. *Dreamscapes of Modernity: Sociotechnical Imaginaries and the Fabrication of Power*. Chicago: University of Chicago Press.

Jenkins, Henry. 2006. *Convergence Culture: Where Old and New Media Collide*. New York: New York University Press.

Jenkins, Henry, Sam Ford, and Joshua Green. 2013. *Spreadable Media: Creating Value and Meaning in a Networked Culture*. New York: New York University Press.

Jenkins, Henry, Sangita Shresthova, Liana Gamber-Thompson, Neta Kligler-Vilenchik and Arely Zimmerman. 2016. *By Any Media Necessary: The New Youth Activism*. New York: New York University Press.

John, Nicholas A. 2013. Sharing and Web 2.0: The emergence of a keyword. *New Media & Society* 15 (2): 167–182. https://doi.org/10.1177/1461444812450684.

John, Nicholas A. 2017. *The Age of Sharing*. Malden, MA: Polity.

John, Nicholas A., and Shira Dvir-Gvirsman. 2015. "I don't like you any more": Facebook unfriending by Israelis during the Israel–Gaza conflict of 2014. *Journal of Communication* 65 (6): 953–974. https://doi.org/10.1111/jcom.12188.

Johns, Adrian. 1998. *The Nature of the Book: Print and Knowledge in the Making*. Chicago: University of Chicago Press.

Johnson, Jessica. 2018. The self-radicalization of white men: "Fake news" and the affective networking of paranoia. *Communication Culture & Critique* 11 (1): 100–115. https://doi.org/10.1093/ccc/tcx014.

Johnson, Thomas, and David D. Perlmutter. 2011. *New Media, Campaigning and the 2008 Facebook Election*. London: Routledge.

Jones, Graham M. 2017. *Magic's Reason: An Anthropology of Analogy*. Chicago: University of Chicago Press.

Kalogeropoulos, Antonis, Samuel Negredo, Ike Picone, and Rasmus Kleis Nielsen. 2017. Who shares and comments on news? A cross-national comparative analysis of online and social media participation. *Social Media + Society* 3 (4): 1–12. https://doi.org/10.1177/2056305117735754.

Kalogeropoulos, Antonis, and Rasmus Kleis Nielsen. 2018. Investing in online video news: A cross-national analysis of news organizations' enterprising approach to digital media. *Journalism Studies* 19 (15): 2207–2224. https://doi.org/10.1080/1461670X.2017.1331709.

Kalsnes, Bente, Arne H. Krumsvik, and Tanja Storsul. 2014. Social media as a political backchannel: Twitter use during televised election debates in Norway. *Aslib Journal of Information Management* 66 (3): 313–328. http://dx.doi.org/10.1108/AJIM-09-2013-0093.

Karapanos, Evangelos, Pedro Teixeira, and Ruben Gouveia. 2016. Need fulfillment and experiences on social media: A case on Facebook and WhatsApp. *Computers in Human Behavior* 55:888–897. https://doi.org/10.1016/j.chb.2015.10.015.

Karim, Karim Haiderali, ed. 2003. *The Media of Diaspora*. London: Routledge.

Katz, James E., and Elizabeth Thomas Crocker. 2015. Selfies and photo messaging as visual conversation: Reports from the United States, United Kingdom and China. *International Journal of Communication* 9:1861–1872.

Kaun, Anne, and Fredrik Stiernstedt. 2014. Facebook time: Technological and institutional affordances for media memories. *New Media & Society* 16 (7): 1154–1168. https://doi.org/10.1177/1461444814544001.

Kavanagh, Barry. 2016. Emoticons as a medium for channeling politeness within American and Japanese online blogging communities. *Language & Communication* 48:53–65. https://doi.org/10.1016/j.langcom.2016.03.003.

Kaye, D. Bondy Valdovinos, Xu Chen, and Jing Zeng. 2021. The co-evolution of two Chinese mobile short video apps: Parallel platformization of Douyin and TikTok. *Mobile Media & Communication* 9 (2): 229–253. https://doi.org/10.1177/2050157920952120.

Khamis, Susie, Lawrence Ang, and Raymond Welling. 2017. Self-branding, "micro-celebrity" and the rise of social media influencers. *Celebrity Studies* 8 (2): 191–208. https://doi.org/10.1080/19392397.2016.1218292.

Kies, Bridget. 2021. Remediating the celebrity roast: The place of mean tweets on late-night television. *Television & New Media* 22 (5): 516–528. https://doi.org/10.1177/1527476419892581.

Kim, Su Jung. 2016. A repertoire approach to cross-platform media use behavior. *New Media & Society* 18 (3): 353–372. https://doi.org/10.1177/1461444814543162.

Kim, Yoojung, Dongyoung Sohn, and Sejung Marina Choi. 2011. Cultural difference in motivations for using social network sites: A comparative study of American and Korean college students. *Computers in Human Behavior* 27 (1): 365–372. https://doi.org/10.1016/j.chb.2010.08.015.

Kim, Young Yun. 2012. Comparing intercultural communication. In *Handbook of Comparative Communication Research*, edited by Frank Esser and Thomas Hanitzsch, 119–133. New York: Routledge.

Kline, Ronald R. 2000. *Consumers in the Country: Technology and Social Change in Rural America*. Baltimore: Johns Hopkins University Press.

Kling, Rob. 1992. Audiences, narratives, and human values in social studies of technology. *Science, Technology, & Human Values* 17 (3): 349–365. https://doi.org/10.1177/016224399201700305.

Kluitenberg, Eric. 2011. On the archeology of imaginary media. In *Media Archaeology: Approaches, Applications, and Implications*, edited by Erkki Huhtamo and Jussi Parikka, 48–69. Berkeley: University of California Press.

Knobel, Michele, and Colin Lankshear. 2008. Remix: The art and craft of endless hybridization. *Journal of Adolescent & Adult Literacy* 52 (1): 22–33. https://doi.org/10.1598/JAAL.52.1.3.

Knorr-Cetina, Karin. 2009. The synthetic situation: Interactionism for a global world. *Symbolic Interaction* 32 (1): 61–87. https://doi.org/10.1525/si.2009.32.1.61.

Koskela, Hille. 2004. Webcams, TV shows and mobile phones: Empowering exhibitionism. *Surveillance & Society* 2 (2/3): 199–215.

Kraidy, Marwan M. 2005. *Hybridity, or the Cultural Logic of Globalization.* Philadelphia: Temple University Press.

Kraidy, Marwan M. 2009. *Reality Television and Arab Politics: Contention in Public Life.* New York: Cambridge University Press.

Kraidy, Marwan M. 2016. *The Naked Blogger of Cairo: Creative Insurgency in the Arab World.* Cambridge, MA: Harvard University Press.

Ksiazek, Thomas B., Limor Peer, and Kevin Lessard. 2016. User engagement with online news: Conceptualizing interactivity and exploring the relationship between online news videos and user comments. *New Media & Society* 18 (3): 502–520. https://doi.org/10.1177/1461444814545073.

Ku, Yi-Cheng, Rui Chen, and Han Zhang. 2013. Why do users continue using social networking sites? An exploratory study of members in the United States and Taiwan. *Information & Management* 50 (7): 571–581. https://doi.org/10.1016/j.im.2013.07.011.

Kwak, Haewoon, Changhyun Lee, Hosung Park, and Sue Moon. 2010. What is Twitter, a social network or a news media? In *Proceedings of the 19th International Conference on World Wide Web,* 591–600. New York: Association for Computing Machinery. https://doi.org/10.1145/1772690.1772751.

Lacour, Philippe, Any Freitas, Aurélien Bénel, Franck Eyraud, and Diana Zambon. 2013. Enhancing linguistic diversity through collaborative translation. In *Social Media and Minority Languages: Convergence and the Creative Industries,* edited by Elin Haf Gruffydd Jones and Enrique Uribe-Jongbloed, 159–172. Bristol, UK: Multilingual Matters.

LaRose, Robert, Regina Connolly, Hyegyu Lee, Kang Li, and Kayla D. Hales. 2014. Connection overload? A cross cultural study of the consequences of social media connection. *Information Systems Management* 31 (1): 59–73. https://doi.org/10.1080/10580530.2014.854097.

Larsson, Anders Olof. 2015. Comparing to prepare: Suggesting ways to study social media today—and tomorrow. *Social Media + Society* 1 (1): 1–2. https://doi.org/10.1177/2056305115578680.

Lasorsa, Dominic L., Seth C. Lewis, and Avery E. Holton. 2012. Normalizing Twitter: Journalism practice in an emerging communication space. *Journalism Studies* 13 (1): 19–36. https://doi.org/10.1080/1461670X.2011.571825.

Latonero, Mark, and Paula Kift. 2018. On digital passages and borders: Refugees and the new infrastructure for movement and control. *Social Media + Society* 4 (1): 1–11. https://doi.org/10.1177/2056305118764432.

Lazer, David, Alex Pentland, Lada Adamic, Sinan Aral, Albert-László Barabási, Devon Brewer, Nicholas Christakis, et al. 2009. Computational social science. *Science (American Association for the Advancement of Science)* 323 (5915): 721–723. https://doi.org /10.1126/science.1167742.

Lécuyer, Christophe. 2006. *Making Silicon Valley: Innovation and the Growth of High Tech, 1930–1970*. Cambridge, MA: MIT Press.

Lee, Jae Kook, Jihyang Choi, Cheonsoo Kim, and Yonghwan Kim. 2014. Social media, network heterogeneity, and opinion polarization. *Journal of Communication* 64 (4): 702–722. https://doi.org/10.1111/jcom.12077.

Lehman-Wilzig, Sam, and Nava Cohen-Avigdor. 2004. The natural life cycle of new media evolution: Inter-media struggle for survival in the internet age. *New Media & Society* 6 (6): 707–730. https://doi.org/10.1177/146144804042524.

Lemke, Jeslyn, and Endalk Chala. 2016. Tweeting democracy: An ethnographic content analysis of social media use in the differing politics of Senegal and Ethiopia's newspapers. *Journal of African Media Studies* 8 (2): 167–185. https://doi.org/10.1386 /jams.8.2.167_1.

Lenihan, Aoife. 2014. Investigating language policy in social media: Translation practices on Facebook. In *The Language of Social Media: Identity and Community on the Internet*, edited by Philip Seargeant and Caroline Tagg, 208–227. Houndmills, UK: Palgrave Macmillan.

Leppänen, Sirpa, and Ari Häkkinen. 2013. Buffalaxed superdiversity: Representations of the other on YouTube. *Diversities* 14 (2): 17–33. https://unesdoc.unesco.org /ark:/48223/pf0000222415.

Leurs, Koen. 2019. Transnational connectivity and the affective paradoxes of digital care labour: Exploring how young refugees technologically mediate co-presence. *European Journal of Communication* 34 (6): 641–649. https://doi.org/10.1177/0267323119886166.

Leurs, Koen, and Madhuri Prabhakar. 2018. Doing digital migration studies: Methodological considerations for an emerging research focus. In *Qualitative Research in European Migration Studies*, edited by Ricard Zapata-Barrero and Evren Yalaz, 247–266. Cham, Switzerland: Springer International.

Leurs, Koen, and Kevin Smets. 2018. Five questions for digital migration studies: Learning from digital connectivity and forced migration in(to) Europe. *Social Media + Society* 4 (1): 1–16. https://doi.org/10.1177/2056305118764425.

Levy, Mark R., and Michael Gurevitch. 1993. Editor's Note. *Journal of Communication* 43 (3): 4–5. https://doi.org/10.1111/j.1460-2466.1993.tb01270.x.

Lewis, Rebecca. 2020. "This is what the news won't show you": YouTube creators and the reactionary politics of micro-celebrity. *Television & New Media* 21 (2): 201–217. https://doi.org/10.1177/1527476419879919.

Liao, Q. Vera, and Wai-Tat Fu. 2013. Beyond the filter bubble: Interactive effects of perceived threat and topic involvement on selective exposure to information. In *Proceedings of the SIGCHI Conference on Human Factors in Computing Systems*, 2359–2368. New York: Association for Computing Machinery. https://doi.org/10.1145/2470654.2481326.

Lievrouw, Leah A., and Sonia M. Livingstone. 2002. Introduction: The Social Shaping and Consequences of ICTs. In *Handbook of New Media: Social Shaping and Consequences of ICTs*, edited by Leah A. Lievrouw and Sonia Livingstone, 1–16. London: SAGE.

Lijphart, Arend. 1971. Comparative politics and the comparative method. *The American Political Science Review* 65 (3): 682–693. https://www.jstor.org/stable/1955513.

Lim, Tae-Seop. 2002. Language and verbal communication across cultures. In *Handbook of International and Intercultural Communication*, edited by William B. Gudykunst and Bella Mody, 69–87. Thousand Oaks, CA: SAGE.

Lin, Luc Chia-Shin. 2016. Convergence in election campaigns: The frame contest between Facebook and mass media. *Convergence* 22 (2): 199–214. https://doi.org/10.1177/1354856514545706.

Lin, Trisha T. C. 2019. Why do people watch multiscreen videos and use dual screening? Investigating users' polychronicity, media multitasking motivation, and media repertoire. *International Journal of Human-Computer Interaction* 35 (18): 1672–1680. https://doi.org/10.1080/10447318.2018.1561813.

Ling, Rich. 2020. Confirmation bias in the era of mobile news consumption: The social and psychological dimensions. *Digital Journalism* 8 (5): 596–604. https://doi.org/10.1080/21670811.2020.1766987.

Litt, Eden. 2012. *Knock, knock*. Who's there? The imagined audience. *Journal of Broadcasting & Electronic Media* 56 (3): 330–345. https://doi.org/10.1080/08838151.2012.705195.

Litt, Eden, and Eszter Hargittai. 2016. The imagined audience on social network sites. *Social Media + Society* 2 (1): 1–12. https://doi.org/10.1177/2056305116633482.

Livingstone, Sonia. 2003. On the challenges of cross-national comparative media research. *European Journal of Communication* 18 (4): 477–500. https://doi.org/10.1177/0267323103184003.

Livingstone, Sonia. 2012. Challenges to comparative research in a globalizing media landscape. In *Handbook of Comparative Communication Research*, edited by Frank Esser and Thomas Hanitzsch, 415–429. New York: Routledge.

Lobato, Ramon. 2018. Rethinking international TV flows research in the age of Netflix. *Television & New Media* 19 (3): 241–256. https://doi.org/10.1177/1527476417708245.

Long, Pamela O. 2015. Trading zones in early modern Europe. *Isis* 106 (4): 840–847. https://doi.org/10.1086/684652.

Lotan, Gilad, Erhardt Graeff, Mike Ananny, Devin Gaffney, and Ian Pearce. 2011. The revolutions were tweeted: Information flows during the 2011 Tunisian and Egyptian revolutions. *International Journal of Communication* 5:1375–1405.

Lu, Jessica H, and Catherine Knight Steele. 2019. "Joy is resistance": Cross-platform resilience and (re)invention of Black oral culture online. *Information, Communication & Society* 22 (6): 823–837. https://doi.org/10.1080/1369118X.2019.1575449.

Lukito, Josephine. 2020. Coordinating a multi-platform disinformation campaign: Internet research agency activity on three U.S. social media platforms, 2015 to 2017. *Political Communication* 37 (2): 238–255. https://doi.org/10.1080/10584609.2019.1661889.

Madianou, Mirca. 2015. Polymedia and ethnography: Understanding the social in social media. *Social Media + Society* 1 (1): 1–3. https://doi.org/10.1177/2056305115578675.

Madianou, Mirca. 2016. Ambient co-presence: Transnational family practices in polymedia environments. *Global Networks* 16 (2): 183–201. https://doi.org/10.1111/glob.12105.

Madianou, Mirca. 2019. Technocolonialism: Digital innovation and data practices in the humanitarian response to refugee crises. *Social Media + Society* 5 (3): 1–13. https://doi.org/10.1177/2056305119863146.

Madianou, Mirca, and Daniel Miller. 2012. *Migration and New Media: Transnational Families and Polymedia*. Abingdon, UK: Routledge.

Madianou, Mirca, and Daniel Miller. 2013. Polymedia: Towards a new theory of digital media in interpersonal communication. *International Journal of Cultural Studies* 16 (2): 169–187. https://doi.org/10.1177/1367877912452486.

Mahajan, Khyati, and Samira Shaikh. 2019. Emoji usage across platforms: A case study for the Charlottesville Event. In *WNLP@ ACL*, 160–162. Florence, Italy: Association for Computational Linguistics.

Manovich, Lev. 2002. *The Language of New Media*. Cambridge, MA: MIT Press.

Marcin, Tim. 2020. One man's frustrating journey to recovering his Myspace. *Mashable*, October 10. https://mashable.com/article/how-to-recover-access-myspace-profile.

Mars, Amanda. 2021. Hay que romper eso de que los gringos son dioses . . . No, papi. *El País*, January, 2. https://elpais.com/elpais/2020/12/30/eps/1609327975_051296.html.

Marshall, P. David. 2010. The promotion and presentation of the self: Celebrity as marker of presentational media. *Celebrity Studies* 1 (1): 35–48. https://doi.org/10.1080/19392390903519057.

Marvin, Carolyn. 1988. *When Old Technologies Were New: Thinking About Electric Communication in the Late Nineteenth Century.* New York: Oxford University Press.

Marwick, Alice E. 2013. *Status Update: Celebrity, Publicity, and Branding in the Social Media Age.* New Haven, CT: Yale University Press.

Marwick, Alice E. 2015. Instafame: Luxury selfies in the attention economy. *Public Culture* 27 (1): 137–160. https://doi.org/10.1215/08992363-2798379.

Marwick, Alice E. 2018. Silicon Valley and the social media industry. In *The SAGE Handbook of Social Media,* edited by Jean Burgess, Alice E. Marwick, and Thomas Poell, 314–329. Los Angeles: SAGE.

Marwick, Alice E., and danah boyd. 2011. I tweet honestly, I tweet passionately: Twitter users, context collapse, and the imagined audience. *New Media & Society* 13 (1): 114–133. https://doi.org/10.1177/1461444810365313.

Marwick, Alice E., and danah boyd. 2014. Networked privacy: How teenagers negotiate context in social media. *New Media & Society* 16 (7): 1051–1067. https://doi.org/10.1177/1461444814543995.

Marx, Leo. 1964. *The Machine in the Garden: Technology and the Pastoral Ideal in America.* New York: Oxford University Press.

Marx, Leo, and Merritt Roe Smith. 1994. *Does Technology Drive History? The Dilemma of Technological Determinism.* Cambridge, MA: MIT Press.

Masullo, Gina, M., Paromita Pain, Victoria Y. Chen, Madlin Mekelburg, Nina Springer, and Franziska Troger. 2020. "You really have to have a thick skin": A cross-cultural perspective on how online harassment influences female journalists. *Journalism* 21 (7): 877–895. https://doi.org/10.1177/1464884918768500.

Matamoros-Fernández, Ariadna. 2017. Platformed racism: The mediation and circulation of an Australian race-based controversy on Twitter, Facebook and YouTube. *Information, Communication & Society* 20 (6): 930–946. https://doi.org/10.1080/1369118X.2017.1293130.

Matassi, Mora, and Pablo Boczkowski. 2021. An agenda for comparative social media studies: The value of understanding social media practices from cross-media, cross-national, and cross-platform perspectives. *International Journal of Communication* 15:207–228. https://ijoc.org/index.php/ijoc/article/view/15042.

Matassi, Mora, Eugenia Mitchelstein, and Pablo J. Boczkowski. 2022. Social media repertoires: Social structure and platform use. *The Information Society* 38 (2): 133–146. https://doi.org/10.1080/01972243.2022.2028208.

McCombs, Maxwell E., and Donald L. Shaw. 1972. The agenda-setting function of mass media. *Public Opinion Quarterly* 36 (2): 176–187. https://doi.org/10.1086/267990.

McCombs, Maxwell E., Donald L. Shaw, and David H. Weaver. 2014. New directions in agenda-setting theory and research. *Mass Communication and Society* 17 (6): 781–802. https://doi.org/10.1080/15205436.2014.964871.

McCombs, Maxwell, and Sebastián Valenzuela. 2021. *Setting the Agenda: Mass Media and Public Opinion*. 3rd edition. Cambridge: Polity.

McLelland, Jack, Haiqing Yu, and Gerard Goggin. 2018. Alternative histories of social media in Japan and China. In *The SAGE Handbook of Social Media*, edited by Jean Burgess, Alice E. Marwick, and Thomas Poell, 53–68. Los Angeles: SAGE.

McLeod, Douglas M., and James K. Hertog. 1992. The manufacture of "public opinion" by reporters: Informal cues for public perceptions of protest groups. *Discourse & Society* 3 (3): 259–275. https://doi.org/10.1177/0957926592003003001.

McLuhan, Marshall. (1964) 2003. *Understanding Media: The Extensions of Man*. New York: Ginko Press.

McPherson, Miller, Lynn Smith-Lovin, and James M. Cook. 2001. Birds of a feather: Homophily in social networks. *Annual Review of Sociology* 27 (1): 415–444. https://doi.org/10.1146/annurev.soc.27.1.415.

Meehan, Mary Beth, and Fred Turner. 2021. *Seeing Silicon Valley: Life Inside a Fraying America*. Chicago: University of Chicago Press.

Mellado, Claudia, and Alfred Hermida. 2021. The promoter, celebrity, and joker roles in journalists' social media performance. *Social Media + Society* 7 (1): 1–11. https://doi.org/10.1177/2056305121990643.

Messing, Solomon, and Sean J. Westwood. 2014. Selective exposure in the age of social media: Endorsements trump partisan source affiliation when selecting news online. *Communication Research* 41 (8): 1042–1063. https://doi.org/10.1177/0093650212466406.

Meyrowitz, Joshua. 1985. *No Sense of Place: The Impact of Electronic Media on Social Behavior*. New York: Oxford University Press.

Miller, Daniel, Jolynna Sinanan, Xinyuan Wang, Tom McDonald, Nell Haynes, Elisabetta Costa, Juliano Spyer, et al. 2016. *How the World Changed Social Media*. London: University College London Press.

Mocanu, Delia, Andrea Baronchelli, Nicola Perra, Bruno Gonçalves, Qian Zhang, and Alessandro Vespignani. 2013. The Twitter of Babel: Mapping world languages through microblogging platforms. *PLoS ONE* 8 (4): e61981. https://doi.org/10.1371/journal.pone.0061981.

Morley, David, and Kevin Robins. 2002. *Spaces of Identity: Global Media, Electronic Landscapes and Cultural Boundaries*. Abingdon, UK: Routledge.

Morris, Nancy, and Silvio Waisbord, eds. 2001. *Media and Globalization: Why the State Matters*. Lanham, MD: Rowman & Littlefield.

Mosca, Lorenzo, and Mario Quaranta. 2016. News diets, social media use and non-institutional participation in three communication ecologies: Comparing Germany, Italy and the UK. *Information, Communication & Society* 19 (3): 325–345. https://doi .org/10.1080/1369118X.2015.1105276.

Mourão, Rachel R. 2019. From mass to elite protests: News coverage and the evolution of antigovernment demonstrations in Brazil. *Mass Communication and Society* 22 (1): 49–71. https://doi.org/10.1080/15205436.2018.1498899.

Mullaney, Thomas S. 2017. *The Chinese Typewriter: A History*. Cambridge, MA: MIT Press.

Murray, Susan, and Laurie Ouellette. 2004. *Reality TV: Remaking Television Culture*. New York: New York University Press.

Murthy, Dhiraj. 2018. *Twitter: Social Communication in the Twitter Age*. Cambridge: Polity.

Nakamura, Lisa, and Peter Chow-White, eds. 2012. *Race After the Internet*. New York: Routledge.

Nedelcu, Mihaela. 2012. Migrants' new transnational habitus: Rethinking migration through a cosmopolitan lens in the digital age. *Journal of Ethnic and Migration Studies* 38 (9): 1339–1356. https://doi.org/10.1080/1369183X.2012.698203.

Nedelcu, Mihaela, and Malika Wyss. 2016. "Doing family" through ICT-mediated ordinary co-presence: Transnational communication practices of Romanian migrants in Switzerland. *Global Networks* 16 (2): 202–218. https://doi.org/10.1111 /glob.12110.

Nee, Rebecca C., and Valerie Barker. 2020. Co-viewing virtually: Social outcomes of second screening with televised and streamed content. *Television & New Media* 21 (7): 712–729. https://journals.sagepub.com/doi/abs/10.1177/1527476419853450.

Neuman, W. Russell, Lauren Guggenheim, S. Mo Jang, and Soo Young Bae. 2014. The dynamics of public attention: Agenda-setting theory meets big data. *Journal of Communication* 64 (2): 193–214. https://doi.org/10.1111/jcom.12088.

Nielsen, Rasmus Kleis, and Sarah Anne Ganter. 2022. *The Power of Platforms: Shaping Media and Society*. New York: Oxford University Press.

Nielsen, Rasmus Kleis, and Kim Christian Schrøder. 2014. The relative importance of social media for accessing, finding, and engaging with news: An eight-country cross-media comparison. *Digital Journalism* 2 (4): 472–489. https://doi.org/10.1080/21670811.2013.872420.

Nissenbaum, Asaf, and Limor Shifman. 2022. Laughing alone, together: Local user-generated satirical responses to a global event. *Information, Communication & Society* 25 (7): 924–941. https://doi.org/10.1080/1369118X.2020.1804979.

Noble, Safiya Umoja. 2018. *Algorithms of Oppression: How Search Engines Reinforce Racism*. New York: New York University Press.

Norris, Pippa. 2009. Comparative political communications: Common frameworks or Babelian confusion? *Government and Opposition* 44 (3): 321–340. https://doi.org/10.1111/j.1477-7053.2009.01290.x.

Ong, Walter J. 1982. *Orality and Literacy: The Technologizing of the Word*. London: Methuen.

Orben, Amy. 2020. The Sisyphean cycle of technology panics. *Perspectives on Psychological Science* 15 (5): 1143–1157. https://doi.org/10.1177/1745691620919372.

Ozimek, Phillip, and Hans-Werner Bierhoff. 2020. All my online-friends are better than me—Three studies about ability-based comparative social media use, self-esteem, and depressive tendencies. *Behaviour & Information Technology* 39 (10): 1110–1123. https://doi.org/10.1080/0144929X.2019.1642385.

Papa, Venetia, and Dimitra L. Milioni. 2016. "I don't wear blinkers, all right?": The multiple meanings of civic identity in the *Indignados* and the role of social media. *Javnost-The Public* 23 (3): 290–306. https://doi.org/10.1080/13183222.2016.1210464.

Papacharissi, Zizi. 2009. The virtual geographies of social networks: A comparative analysis of Facebook, LinkedIn and ASmallWorld. *New Media & Society* 11 (1–2): 199–220. https://doi.org/10.1177/1461444808099577.

Papacharissi, Zizi. 2010. *A Private Sphere: Democracy in a Digital Age*. Cambridge: Polity.

Papacharissi, Zizi, and Maria de Fatima Oliveira. 2012. Affective news and networked publics: The rhythms of news storytelling on #Egypt. *Journal of Communication* 62 (2): 266–282. https://doi.org/10.1111/j.1460-2466.2012.01630.x.

Parikka, Jussi. 2012. *What Is Media Archaeology?* Cambridge: Polity.

Pariser, Eli. 2011. *The Filter Bubble: What the Internet Is Hiding from You*. New York: Penguin.

Parisi, Lorenza, and Francesca Comunello. 2020. Dating in the time of "relational filter bubbles": Exploring imaginaries, perceptions and tactics of Italian dating app users. *The Communication Review* 23 (1): 66–89. https://doi.org/10.1080/10714421.2019.1704111.

Paulussen, Steve, and Raymond A. Harder. 2014. Social media references in newspapers: Facebook, Twitter and YouTube as sources in newspaper journalism. *Journalism Practice* 8 (5): 542–551. https://doi.org/10.1080/17512786.2014.894327.

Pearce, Warren, Suay M. Özkula, Amanda K. Greene, Lauren Teeling, Jennifer S. Bansard, Janna Joceli Omena, and Elaine Teixeira Rabello. 2020. Visual cross-platform analysis: Digital methods to research social media images. *Information, Communication & Society* 23 (2): 161–180. https://doi.org/10.1080/1369118X.2018.1486871.

Peirce, Charles Sanders. 1984. Some consequences of four incapacities. In *Writings of Charles Sanders Peirce, A Chronological Edition. Vol. 2, 1867–1871*, 211–242. Bloomington: Indiana University Press.

Perse, Elizabeth M. 1990. Audience selectivity and involvement in the newer media environment. *Communication Research* 17 (5): 675–697.

Peters, Benjamin. 2009. And lead us not into thinking the new is new: A bibliographic case for new media history. *New Media & Society* 11 (1–2): 13–30. https://doi.org/10.1177/1461444808099572.

Pfau, Michael. 2008. Epistemological and disciplinary intersections. *Journal of Communication* 58 (4): 597–602. https://doi.org/10.1111/j.1460-2466.2008.00414.x.

Pinch, Trevor J., and Wiebe E. Bijker. 1984. The social construction of facts and artefacts: Or how the sociology of science and the sociology of technology might benefit each other. *Social Studies of Science* 14 (3): 399–441. https://doi.org/10.1177/030631284014003004.

Pittman, Matthew, and Brandon Reich. 2016. Social media and loneliness: Why an Instagram picture may be worth more than a thousand Twitter words. *Computers in Human Behavior* 62:155–167. https://doi.org/10.1016/j.chb.2016.03.084.

Plantin, Jean-Christophe, Carl Lagoze, Paul N. Edwards, and Christian Sandvig. 2018. Infrastructure studies meet platform studies in the age of Google and Facebook. *New Media & Society* 20 (1): 293–310. https://doi.org/10.1177/1461444816661553.

Plotnick, Rachel. 2018. *Power Button: A History of Pleasure, Panic, and the Politics of Pushing*. Cambridge, MA: MIT Press.

Polyakova, Alina. 2019. What the Mueller report tells us about Russian influence operations. *Brookings*, April 18. https://www.brookings.edu/blog/order-from-chaos/2019/04/18/what-the-mueller-report-tells-us-about-russian-influence-operations.

Postill, John. 2014. Democracy in an age of viral reality: A media epidemiography of Spain's indignados movement. *Ethnography* 15 (1): 51–69. https://doi.org/10.1177/1466138113502513.

Postman, Neil. 1986. *Amusing Ourselves to Death: Public Discourse in the Age of Show Business*. New York: Penguin Books.

Prior, Markus. 2005. News vs. entertainment: How increasing media choice widens gaps in political knowledge and turnout. *American Journal of Political Science* 49 (3): 577–592. https://doi.org/10.1111/j.1540-5907.2005.00143.x.

Psarras, Evie. 2020. "It's a mix of authenticity and complete fabrication" Emotional camping: The cross-platform labor of the *Real Housewives. New Media & Society.* https://doi.org/10.1177/1461444820975025.

Qiu, Lin, Han Lin, and Angela K.-y. Leung. 2013. Cultural differences and switching of in-group sharing behavior between an American (Facebook) and a Chinese (Renren) social networking site. *Journal of Cross-Cultural Psychology* 44 (1): 106–121. https://doi.org/10.1177/0022022111434597.

Quan-Haase, Anabel, and Alyson L. Young. 2010. Uses and gratifications of social media: A comparison of Facebook and instant messaging. *Bulletin of Science, Technology & Society* 30 (5): 350–361. https://doi.org/10.1177/0270467610380009.

Rainie, Harrison, and Barry Wellman. 2012. *Networked: The New Social Operating System.* Cambridge, MA: MIT Press.

Renninger, Bryce J. 2015. "Where I can be myself . . . where I can speak my mind": Networked counterpublics in a polymedia environment. *New Media & Society* 17 (9): 1513–1529. https://doi.org/10.1177/1461444814530095.

Rice, Ronald E. 1999. Artifacts and paradoxes in new media. *New Media & Society* 1 (1): 24–32. https://doi.org/10.1177/1461444899001001005.

Roberts, Sarah T. 2019. *Behind the Screen: Content Moderation in the Shadows of Social Media.* New Haven, CT: Yale University Press.

Rymarczuk, Robin. 2016. Same old story: On non-use and resistance to the telephone and social media. *Technology in Society* 45:40–47. https://doi.org/10.1016/j.techsoc.2016.02.003.

Salameh, Mohammad, Saif Mohammad, and Svetlana Kiritchenko. 2015. Sentiment after translation: A case-study on Arabic social media posts. In *Proceedings of the 2015 Conference of the North American Chapter of the Association for Computational Linguistics: Human Language Technologies,* 767–777. Denver, CO: Association for Computational Linguistics.

Saldaña, Magdalena, Shannon C. McGregor, and Homero Gil de Zúñiga. 2015. Social media as a public space for politics: Cross-national comparison of news consumption and participatory behaviors in the United States and the United Kingdom. *International Journal of Communication* 9 (1): 3304–3326.

Sánchez-Querubín, Natalia, and Richard Rogers. 2018. Connected routes: Migration studies with digital devices and platforms. *Social Media + Society* 4 (1): 1–13. https://doi.org/10.1177/2056305118764427.

Saunders, Jessica F., and Asia A. Eaton. 2018. Snaps, selfies, and shares: How three popular social media platforms contribute to the sociocultural model of disordered eating among young women. *Cyberpsychology, Behavior, and Social Networking* 21 (6): 343–354. https://doi.org/10.1089/cyber.2017.0713.

Saxenian, AnnaLee. 1996. *Regional Advantage: Culture and Competition in Silicon Valley and Route 128*. Cambridge, MA: Harvard University Press.

Sayre, Ben, Leticia Bode, Dhavan Shah, Dave Wilcox, and Chirag Shah. 2010. Agenda setting in a digital age: Tracking attention to California Proposition 8 in social media, online news and conventional news. *Policy & Internet* 2 (2): 7–32. https://doi.org/10.2202/1944-2866.1040.

Schmidt, Jan-Hinrik, Lisa Merten, Uwe Hasebrink, Isabelle Petrich, and Amelie Rolfs. 2019. How do intermediaries shape news-related media repertoires and practices? Findings from a qualitative study. *International Journal of Communication* 13:853–873.

Schmitz Weiss, Amy. 2015. The digital and social media journalist: A comparative analysis of journalists in Argentina, Brazil, Colombia, Mexico, and Peru. *International Communication Gazette* 77 (1): 74–101. https://doi.org/10.1177/17480485 14556985.

Schrøder, Kim Christian. 2015. News media old and new: Fluctuating audiences, news repertoires and locations of consumption. *Journalism Studies* 16 (1): 60–78. https://doi.org/10.1080/1461670X.2014.890332.

Schroeder, Ralph. 2016. The globalization of on-screen sociability: Social media and tethered togetherness. *International Journal of Communication* 10:5626–5643.

Schultz, Friederike, Sonja Utz, and Anja Göritz. 2011. Is the medium the message? Perceptions of and reactions to crisis communication via Twitter, blogs and traditional media. *Public Relations Review* 37 (1): 20–27. https://doi.org/10.1016/j.pubrev .2010.12.001.

Schünemann, Wolf J. 2020. Ready for the world? Measuring the (trans-)national quality of political issue publics on Twitter. *Media and Communication* 8 (4): 40–52. http://dx.doi.org/10.17645/mac.v8i4.3162.

Scolari, Carlos A. 2009. Transmedia storytelling: Implicit consumers, narrative worlds, and branding in contemporary media production. *International Journal of Communication* 3:586–606.

Scolari, Carlos A. 2012. Media ecology: Exploring the metaphor to expand the theory. *Communication Theory* 22 (2): 204–225. https://doi.org/10.1111/j.1468-2885 .2012.01404.x.

Scolari, Carlos A. 2013. Media evolution: Emergence, dominance, survival and extinction in the media ecology. *International Journal of Communication* 7:1418–1441.

Scolere, Leah, Urszula Pruchniewska, and Brooke Erin Duffy. 2018. Constructing the platform-specific self-brand: The labor of social media promotion. *Social Media + Society* 4 (3): 1–11. https://doi.org/10.1177/2056305118784768.

Seagoe, May V. 1951. Children's television habits and preferences. *The Quarterly of Film, Radio, and Television* 6 (2): 143–153. https://doi.org/10.2307/1209900.

Seidman, Steven A. 2008. *Posters, Propaganda, and Persuasion in Election Campaigns around the World and through History*. New York: Peter Lang.

Selva, Donatella. 2016. Social television: Audience and political engagement. *Television & New Media* 17 (2): 159–173. https://doi.org/10.1177/1527476415616192.

Seo, Hyunjin, and Robert Faris. 2021. Special section on comparative approaches to mis/disinformation. *International Journal of Communication* 15:1165–1172.

Shahin, Saif, Junki Nakahara, and Mariana Sánchez. 2021. Black Lives Matter goes global: Connective action meets cultural hybridity in Brazil, India, and Japan. *New Media & Society*. https://doi.org/10.1177/14614448211057106.

Shane-Simpson, Christina, Adriana Manago, Naomi Gaggi, and Kristen Gillespie-Lynch. 2018. Why do college students prefer Facebook, Twitter, or Instagram? Site affordances, tensions between privacy and self-expression, and implications for social capital. *Computers in Human Behavior* 86:276–288. https://doi.org/10.1016/j.chb.2018.04.041.

Shifman, Limor. 2007. Humor in the age of digital reproduction: Continuity and change in internet-based comic texts. *International Journal of Communication* 1:187–209.

Sibilia, Paula. 2008. *La Intimidad Como Espectáculo*. Buenos Aires: Fondo de Cultura Económica.

Siles, Ignacio. 2017. *Networked Selves: Trajectories of Blogging in the United States and France*. New York: Peter Lang.

Singh, Manish. 2021. Facebook, Twitter, WhatsApp face tougher rules in India. *Tech Crunch*, February 25. https://techcrunch.com/2021/02/25/india-announces-sweeping-guidelines-for-social-media-on-demand-streaming-firms-and-digital-news-outlets.

Sisto, Davide. 2020. *Online Afterlives: Immortality, Memory, and Grief in Digital Culture*. Cambridge, MA: MIT Press.

Skoric, Marko M., and Nathaniel Poor. 2013. Youth engagement in Singapore: The interplay of social and traditional media. *Journal of Broadcasting & Electronic Media* 57 (2): 187–204. https://doi.org/10.1080/08838151.2013.787076.

Skoric, Marko M., Qinfeng Zhu, and Jih-Hsuan Tammy Lin. 2018. What predicts selective avoidance on social media? A study of political unfriending in Hong Kong and Taiwan. *American Behavioral Scientist* 62 (8): 1097–1115. https://doi.org/10.1177/0002764218764251.

Slater, Michael D. 2007. Reinforcing spirals: The mutual influence of media selectivity and media effects and their impact on individual behavior and social identity. *Communication Theory* 17 (3): 281–303. https://doi.org/10.1111/j.1468-2885.2007.00296.x.

Smoliarova, Anna S., Tamara M. Gromova, and Natalia A. Pavlushkina. 2018. Emotional stimuli in social media user behavior: Emoji reactions on a news media Facebook page. In *International Conference on Internet Science*, 242–256. Cham, Switzerland: Springer International.

Snell-Hornby, Mary. 1999. Communicating in the global village: On language, translation and cultural identity. *Current Issues in Language & Society* 6 (2): 103–120. https://doi.org/10.1080/13520529909615539.

So, Clement Y. K. 2010. The rise of Asian communication research: A citation study of SSCI journals. *Asian Journal of Communication* 20 (2): 230–247. https://doi.org/10.1080/01292981003693419.

Solaris. 2020. Capítulo once: Viralidad. *Podium Podcast.* https://www.podiumpodcast.com/podcasts/solaris-podium-os/episodio/3097589/.

Sparks, Colin. 2008. Media systems in transition: Poland, Russia, China. *Chinese Journal of Communication* 1 (1): 7–24. https://doi.org/10.1080/17544750701861871.

Stefanone, Michael A., and Derek Lackaff. 2009. Reality television as a model for online behavior: Blogging, photo, and video sharing. *Journal of Computer-Mediated Communication* 14 (4): 964–987. https://doi.org/10.1111/j.1083-6101.2009.01477.x.

Stefanone, Michael A., Derek Lackaff, and Devan Rosen. 2010. The relationship between traditional mass media and "social media": Reality television as a model for social network site behavior. *Journal of Broadcasting & Electronic Media* 54 (3): 508–525. https://doi.org/10.1080/08838151.2010.498851.

Steinberg, Marc. 2020. LINE as Super App: Platformization in East Asia. *Social Media + Society* 6 (2): 1–10. https://doi.org/10.1177/2056305120933285.

Sterne, Jonathan. 2003. *The Audible Past: Cultural Origins of Sound Reproduction.* Durham, NC: Duke University Press.

Stieglitz, Stefan, and Linh Dang-Xuan. 2013. Emotions and information diffusion in social media—Sentiment of microblogs and sharing behavior. *Journal of Management Information Systems* 29 (4): 217–248. https://doi.org/10.2753/MIS0742-1222290408.

Strangelove, Michael. 2010. *Watching YouTube: Extraordinary Videos by Ordinary People.* Toronto: University of Toronto Press.

Strate, Lance. 2004. A media ecology review. *Communication Research Trends* 23 (2): 3–48.

Straubhaar, Joseph D. 2007. *World Television: From Global to Local.* Thousand Oaks, CA: SAGE.

Streeter, Thomas. 2011. *The Net Effect: Romanticism, Capitalism, and the Internet.* New York: New York University Press.

Su, Chunmeizi. 2019. "Changing dynamics of digital entertainment media in China." PhD diss., Queensland University of Technology.

Sumner, Erin M., Rebecca A. Hayes, Caleb T. Carr, and Donghee Yvette Wohn. 2020. Assessing the cognitive and communicative properties of Facebook Reactions and Likes as lightweight feedback cues. *First Monday* 25 (2). https://doi.org/10.5210/fm.v25i2.9621.

Sundar, S. Shyam. 2008. The MAIN model: A heuristic approach to understanding technology effects on credibility. In *Digital Media, Youth, and Credibility*, edited by Miriam J. Metzger and Andrew J. Flanagin. Cambridge, MA: MIT Press.

Sunstein, Cass R. 2009. *Republic.com 2.0.* Princeton, NJ: Princeton University Press.

Sunstein, Cass R. 2017. *#Republic: Divided Democracy in the Age of Social Media.* Princeton, NJ: Princeton University Press.

Susser, Daniel, Beate Roessler, and Helen Nissenbaum. 2019. Technology, autonomy, and manipulation. *Internet Policy Review* 8 (2). https://doi.org/10.14763/2019.2.1410.

Suzina, Ana Cristina. 2021. English as *lingua franca*. Or the sterilisation of scientific work. *Media, Culture & Society* 43 (1): 171–179. https://doi.org/10.1177/0163443720957906.

Swart, Joëlle, Chris Peters, and Marcel Broersma. 2017. Navigating cross-media news use: Media repertoires and the value of news in everyday life. *Journalism Studies* 18 (11): 1343–1362. https://doi.org/10.1080/1461670X.2015.1129285.

Syvertsen, Trine. 2017. *Media Resistance Protest, Dislike, Abstention.* Cham, Switzerland: Springer International.

Tandoc, Edson C., Jr., Chen Lou, and Velyn Lee Hui Min. 2019. Platform-swinging in a poly-social-media context: How and why users navigate multiple social media platforms. *Journal of Computer-Mediated Communication* 24 (1): 21–35. https://doi.org/10.1093/jcmc/zmy022.

Taneja, Harsh, James G. Webster, Edward C. Malthouse, and Thomas B. Ksiazek. 2012. Media consumption across platforms: Identifying user-defined repertoires. *New Media & Society* 14 (6): 951–968. https://doi.org/10.1177/1461444811436146.

Taylor, T. L. 2018. *Watch Me Play: Twitch and the Rise of Game Live Streaming.* Princeton, NJ: Princeton University Press.

Tenenboim-Weinblatt, Keren, and Chul-joo Lee. 2020. Speaking across communication subfields. *Journal of Communication* 70 (3): 303–309. https://doi.org/10.1093/joc/jqaa012.

Teune, Henry, and Adam Przeworski. 1970. *The Logic of Comparative Social Inquiry.* New York: Wiley-Interscience.

Theocharis, Yannis, Will Lowe, Jan W. Van Deth, and Gema García-Albacete. 2015. Using Twitter to mobilize protest action: Online mobilization patterns and action repertoires in the Occupy Wall Street, Indignados, and Aganaktismenoi movements. *Information, Communication & Society* 18 (2): 202–220. https://doi.org/10.1080/1369118X.2014.948035.

Thompson, Emily Ann. 2002. *The Soundscape of Modernity: Architectural Acoustics and the Culture of Listening in America, 1900–1933.* Cambridge, MA: MIT Press.

Thorburn, David, and Henry Jenkins. 2003. *Rethinking Media Change: The Aesthetics of Transition.* Cambridge, MA: MIT Press.

Thurlow, Crispin. 2018. Digital discourse: Locating language in new/social media. In *The SAGE Handbook of Social Media*, edited by Jean Burgess, Alice E. Marwick, and Thomas Poell, 135–145. Los Angeles: SAGE.

Thurlow, Crispin, and Kristine Mroczek, eds. 2011. *Digital Discourse: Language in the New Media.* New York: Oxford University Press.

Torkjazi, Mojtaba, Reza Rejaie, and Walter Willinger. 2009. Hot today, gone tomorrow: On the migration of MySpace users. In *Proceedings of the 2nd ACM Workshop on Online Social Networks*, 43–48. New York: ACM. https://doi.org/10.1145/1592665.1592676.

Törnberg, Anton, and Petter Törnberg. 2016. Muslims in social media discourse: Combining topic modeling and critical discourse analysis. *Discourse, Context & Media* 13:132–142. https://doi.org/10.1016/j.dcm.2016.04.003.

Trepte, Sabine, Leonard Reinecke, Nicole B. Ellison, Oliver Quiring, Mike Z. Yao, and Marc Ziegele. 2017. A cross-cultural perspective on the privacy calculus. *Social Media + Society* 3 (1): 1–13. https://doi.org/10.1177/2056305116688035.

Trevisan, Filippo. 2020. "Do you want to be a well-informed citizen, or do you want to be sane?": Social media, disability, mental health, and political marginality. *Social Media + Society* 6 (1): 1–11. https://doi.org/10.1177/2056305120913909.

Tufekci, Zeynep. 2014. Big questions for social media big data: Representativeness, validity and other methodological pitfalls. In *Proceedings of the Eighth International*

AAAI Conference on Weblogs and Social Media, 505–514. Palo Alto, CA: The AAAI Press.

Tufekci, Zeynep. 2018. *Twitter and Tear Gas: The Power and Fragility of Networked Protest*. New Haven, CT: Yale University Press.

Tufekci, Zeynep, and Christopher Wilson. 2012. Social media and the decision to participate in political protest: Observations from Tahrir Square. *Journal of Communication* 62 (2): 363–379. https://doi.org/10.1111/j.1460-2466.2012.01629.x.

Turner, Fred. 2006. *From Counterculture to Cyberculture: Stewart Brand, the Whole Earth Network, and the Rise of Digital Utopianism*. Chicago: University of Chicago Press.

Twenge, Jean M., Thomas E. Joiner, Megan L. Rogers, and Gabrielle N. Martin. 2018. Increases in depressive symptoms, suicide-related outcomes, and suicide rates among U.S. adolescents after 2010 and links to increased new media screen time. *Clinical Psychological Science* 6 (1): 3–17. https://doi.org/10.1177/2167702617723376.

Ullmann, Stefanie. 2017. Conceptualising force in the context of the Arab revolutions: A comparative analysis of international mass media reports and Twitter posts. *Discourse & Communication* 11 (2): 160–178. https://doi.org/10.1177/1750481317691859.

Utz, Sonja, Nicole Muscanell, and Cameran Khalid. 2015. Snapchat elicits more jealousy than Facebook: A comparison of Snapchat and Facebook use. *Cyberpsychology, Behavior and Social Networking* 18 (3): 141–146. https://doi.org/10.1089/cyber.2014.0479.

Uy-Tioco, Cecilia. 2007. Overseas Filipino workers and text messaging: Reinventing transnational mothering. *Continuum* 21 (2): 253–265. https://doi.org/10.1080/10304310701269081.

Vaidhyanathan, Siva. 2018. *Antisocial Media: How Facebook Disconnects Us and Undermines Democracy*. New York: Oxford University Press.

Valenzuela, Sebastián, Teresa Correa, and Homero Gil de Zúñiga. 2018. Ties, likes, and tweets: Using strong and weak ties to explain differences in protest participation across Facebook and Twitter use. *Political Communication* 35 (1): 117–134. https://doi.org/10.1080/10584609.2017.1334726.

Valenzuela, Sebastián, Soledad Puente, and Pablo M. Flores. 2017. Comparing disaster news on Twitter and television: An intermedia agenda setting perspective. *Journal of Broadcasting & Electronic Media* 61 (4): 615–637. https://doi.org/10.1080/08838151.2017.1344673.

van Atteveldt, Wouter, and Tai-Quan Peng. 2018. When communication meets computation: Opportunities, challenges, and pitfalls in computational communication science. *Communication Methods and Measures* 12 (2–3): 81–92. https://doi.org/10.1080/19312458.2018.1458084.

Vanden Abeele, Mariek M. P. 2020. Digital wellbeing as a dynamic construct. *Communication Theory* 31 (4): 932–955. https://doi.org/10.1093/ct/qtaa024.

van Dijck, José. 2013. "You have one identity": Performing the self on Facebook and LinkedIn. *Media, Culture & Society* 35 (2): 199–215. https://doi.org/10.1177/01634437 12468605.

van Dijck, José. 2014. Datafication, dataism and dataveillance: Big Data between scientific paradigm and ideology. *Surveillance & Society* 12 (2): 197–208. https://doi.org /10.24908/ss.v12i2.4776.

van Dijck, José. 2020. Governing digital societies: Private platforms, public values. *Computer Law and Security Report* 36:105377. https://doi.org/10.1016/j.clsr.2019 .105377.

van Dijk, Jan. 2006. *The Network Society: Social Aspects of New Media*. Thousand Oaks, CA: SAGE.

Vessey, Rachelle. 2015. Food fight: Conflicting language ideologies in English and French news and social media. *Journal of Multicultural Discourses* 10 (2): 253–271. https://doi.org/10.1080/17447143.2015.1042883.

Vliegenthart, Rens, and Stefaan Walgrave. 2008. The contingency of intermedia agenda setting: A longitudinal study in Belgium. *Journalism & Mass Communication Quarterly* 85 (4): 860–877. https://doi.org/10.1177/107769900808500409.

von Nordheim, Gerret, Karin Boczek, Lars Koppers, and Elena Erdmann. 2018. Digital traces in context: Reuniting a divided public? Tracing the TTIP debate on Twitter and in traditional media. *International Journal of Communication* 12:548–569.

Vraga, Emily K., and Melissa Tully. 2019. Engaging with the other side: Using news media literacy messages to reduce selective exposure and avoidance. *Journal of Information Technology & Politics* 16 (1): 77–86. https://doi.org/10.1080/19331681.2019 .1572565.

Wagner, Claudia, Markus Strohmaier, Alexandra Olteanu, Emre Kıcıman, Noshir Contractor, and Tina Eliassi-Rad. 2021. Measuring algorithmically infused societies. *Nature* (London) 595 (7866): 197–204. https://doi.org/10.1038/s41586-021-03666-1.

Waisbord, S. 2019. *Communication: A Post-Discipline*. Cambridge: Polity.

Waisbord, Silvio. 2020. *El Imperio de la Utopía: Mitos y Realidades de la Sociedad Estadounidense*. Madrid: Ediciones Península.

Walker, Shawn, Dan Mercea, and Marco Bastos. 2019. The disinformation landscape and the lockdown of social platforms. *Information, Communication & Society* 22 (11): 1531–1543. https://doi.org/10.1080/1369118X.2019.1648536.

Walter, Nathan, Michael J. Cody, Sandra J. Ball-Rokeach. 2018. The ebb and flow of communication research: Seven decades of publication trends and research priorities. *Journal of Communication* 68 (2): 424–440. https://doi.org/10.1093/joc/jqx015.

Wartella, Ellen, and Byron Reeves. 1985. Historical trends in research on children and the media: 1900–1960. *Journal of Communication* 35 (2): 118–133. https://doi .org/10.1111/j.1460-2466.1985.tb02238.x.

Weber, Max. 1949. "Objectivity" in social science and social policy. In *The Methodology of the Social Sciences*, 49–112. New York: Free Press.

Webster, James G. 2011. The duality of media: A structurational theory of public attention. *Communication Theory* 21 (1): 43–66. https://doi.org/10.1111/j.1468-2885 .2010.01375.x.

Webster, James G., and Jacob J. Wakshlag. 1983. A theory of television program choice. *Communication Research* 10 (4): 430–446. https://doi.org/10.1177/009365083 010004002.

Wiemann, John M., Suzanne Pingree, and Robert P. Hawkins. 1988. Fragmentation in the field—and the movement toward integration in communication science. *Human Communication Research* 15 (2): 304–310. https://doi.org/10.1111/j.1468-2958.1988 .tb00186.x.

Willems, Wendy. 2014. Provincializing hegemonic histories of media and communication studies: Toward a genealogy of epistemic resistance in Africa. *Communication Theory* 24 (4): 415–434. https://doi.org/10.1111/comt.12043.

Williams, Frederick, Ronald E. Rice, and Everett M. Rogers. 1988. *Research Methods and the New Media*. New York: Free Press.

Winner, Langdon. 1980. Do artifacts have politics? *Daedalus* 109 (1): 121–136. https://www.jstor.org/stable/20024652.

Wohn, Donghee Y., and Eun-Kyung Na. 2011. Tweeting about TV: Sharing television viewing experiences via social media message streams. *First Monday* 16 (3). https://doi.org/10.5210/fm.v16i3.3368.

Wolfsfeld, Gadi, Elad Segev, and Tamir Sheafer. 2013. Social media and the Arab Spring: Politics comes first. *The International Journal of Press/Politics* 18 (2): 115–137. https://doi.org/10.1177/1940161212471716.

World Migration Report. 2020. *Geneva: International Organization for Migration*. https://doi.org/10.18356/b1710e30-en.

Wyatt, Sally M. E. 2003. Non-users also matter: The construction of users and non-users of the Internet. In *How Users Matter: The Co-Construction of Users and*

Technologies, edited by Nelly Oudshoorn and Trevor J. Pinch, 67–79. Cambridge, MA: MIT Press.

Yang, JungHwan, Matthew Barnidge, and Hernando Rojas. 2017. The politics of "Unfriending": User filtration in response to political disagreement on social media. *Computers in Human Behavior* 70:22–29. https://doi.org/10.1016/j.chb.2016.12.079.

Yarchi, Moran, Christian Baden, and Neta Kligler-Vilenchik. 2020. Political polarization on the digital sphere: A cross-platform, over-time analysis of interactional, positional, and affective polarization on social media. *Political Communication* 38 (1–2): 98–139. https://doi.org/10.1080/10584609.2020.1785067.

Zarowsky, Mariano. 2017. *Los Estudios en Comunicación en la Argentina: Ideas, Intelectuales, Tradiciones Político-Culturales*. Buenos Aires: Eudeba.

Zelizer, Barbie. 2016. Communication in the fan of disciplines. *Communication Theory* 26 (3): 213–235. https://doi.org/10.1111/comt.12094.

Zhao, Luolin, and Nicholas John. 2020. The concept of "sharing" in Chinese social media: Origins, transformations and implications. *Information, Communication & Society* 25 (3): 359–375. https://doi.org/10.1080/1369118X.2020.1791216.

Zhao, Xuan, Cliff Lampe, and Nicole B. Ellison. 2016. The social media ecology: User perceptions, strategies and challenges. In *Proceedings of the 2016 CHI Conference on Human Factors in Computing Systems*, 89–100. New York: Association for Computing Machinery. https://doi.org/10.1145/2858036.2858333.

Zhou, Yuchen, Mark Dredze, David A. Broniatowski, and William D. Adler. 2019. Elites and foreign actors among the alt-right: The Gab social media platform. *First Monday* 24 (9). https://doi.org/10.5210/fm.v24i9.10062.

Zhu, Qinfeng, Marko Skoric, and Fei Shen. 2017. I shield myself from thee: Selective avoidance on social media during political protests. *Political Communication* 34 (1): 112–131. https://doi.org/10.1080/10584609.2016.1222471.

Zuboff, Shoshana. 2019. *The Age of Surveillance Capitalism: The Fight for a Human Future at the New Frontier of Power*. New York: Public Affairs.

Index